AF691623

Everyday People Save the Planet and So Can You

Everyday People Save the Planet and So Can You

A Qualitative Examination of Green Lifestyles in Lowcountry South Carolina

Deborah McCarthy Auriffeille

LEXINGTON BOOKS
Lanham • Boulder • New York • London

Published by Lexington Books
An imprint of The Rowman & Littlefield Publishing Group, Inc.
4501 Forbes Boulevard, Suite 200, Lanham, Maryland 20706
www.rowman.com

86-90 Paul Street, London EC2A 4NE

Copyright © 2022 by The Rowman & Littlefield Publishing Group, Inc.

All rights reserved. No part of this book may be reproduced in any form or by any electronic or mechanical means, including information storage and retrieval systems, without written permission from the publisher, except by a reviewer who may quote passages in a review.

British Library Cataloguing in Publication Information Available

Library of Congress Cataloging-in-Publication Data Available

ISBN 978-1-7936-1616-6 (cloth)
ISBN 978-1-7936-1617-3 (electronic)

It is with great love and thankfulness that I dedicate this book to Jean François, Gaspard, and Fin. You three are my one heart.

Contents

Acknowledgments

It is impossible to imagine completing this green living project on my own. During the course of writing this book, I became a mother and raised Gaspard who is now ten years old; moved to a new home; was diagnosed with a herniated disc, heart disease, and lupus; lost Bettie to pancreatic cancer, served as Director of the Urban Studies major; took on the responsibilities of Sociology and Anthropology Departmental Associate Chair; wrote four articles; attended multiple academic conferences to shop my analyses; and completed three new editions of *Environmental Sociology From Analysis to Action* (coedited with Leslie King). My family also evacuated three major hurricanes, endured the devastating thousand-year floods and, more recently, lived through COVID-19, which involved homeschooling our son, quarantining our household four times, and transforming my face-to-face courses to distance education, among other unexpected pandemic-related adventures. The experience has been both, at times, joyful and arduous.

When I think about my own path to writing this book and our shared larger journey to create a more sustainable world, I recollect a recent family ski trip to Beech Mountain, North Carolina. Gaspard had earlier learned to ski, as a five- and six-year-old, in the French Pyrenean mountains where we have family. Due to COVID-19, and other factors, we missed out on later ski trips. Then, in March 2022, we visited Beech Mountain with friends. Gaspard, now ten years old, was excited to show off his skiing skills. However, with a much different body then he wielded as a five-year-old in the Pyrenees, Gaspard found himself flailing all over the mountain while our friends' children zipped down the slopes. On our fourth day, Gaspard wanted to try the Blue trail. He had made progress and I knew he could do it. I also knew what he would see at the top. We disembarked from the lift and skied to the edge. Gaspard looked me straight in the eyes and said, "I'm not doin' it mom." I was in front of him, downslope, and said, firmly, "Yes you are, you can do it." He backed up a little, "no mom." I climbed up a little, "You are doing this, you *can* do this! Just go sidewise." I showed him as I parallel skied to the side lip of the slope. Finally, he inched forward and skied sideways and fell. "Great!" I shouted. "Get up and go back the other way." This went on and on for a torturous amount of time, but my child got down the mountain. We spent the rest of the day mastering Blue. Now, if anyone asks Gaspard about his favorite part of the trip, his answer is always "the Blue trail." He met the mountain head-on and mastered his fears. And he had a joyful and, even, transcendent experience with his family/friends in nature. This story offers a thread that weaves through this entire book. We have a global-sized mountain of a problem and we all have to work together to get down that mountain. Looking over that first precipice is scary, but we have to be brave. We have to look to each other for help. We have to find ways to trust, to innovate, to do things differently, and get down that mountain.

Like Gaspard on the Blue trail, I've backed away from this book many times. I had a lot of excuses, COVID-19, blah, blah, blah. But my loving husband, partner-in-crime and best friend, Jean François, never let me quit. He provided not only emotional strength, but also editing support at key moments of this long-term project. Gaspard's constant refrain "you've got this mom" means absolutely everything to me. Leslie King, of Smith College, read multiple drafts of analyses that fed into this book and acted as a continuous source of thoughtful insight. Since the early days of researching environmental lifestyles, I was assisted by over ten students at the College of Charleston on various activities including: reviewing the literature, interviewing, data cleaning, coding, peer-coding, transcribing, writing, and editing. Most recently, I coauthored with Cat Fleming, a former undergraduate Sociology student, the below- cited *Contexts* article. Cat also "Frannied" Gaspard and became part of our family. She is now pursuing her graduate degree at John Jay College in NYC, and we miss her dearly.

I have also been incredibly lucky to work in a supportive academic community at the College of Charleston. First, I cannot overstate how especially grateful I am for my family-friendly Sociology and Anthropology Department. In terms of financial support, over the full course of carrying out this project, I received over $10,000 in college, school, and departmental grants. I received additional departmental and school funds to attend research conferences. Finally, I have been awarded two sabbaticals to collect and analyze data.

I received generous support from different parts of the Lowcountry community and government. During the bike commuter component of the project, the regional Berkeley, Charleston, and Dorchester Council of Governments (COG) provided funds to hire student researchers and the Planning, Preservation & Sustainability Department of the City of Charleston provided helpful information and maps. The Children's Museum and a host of local libraries and community centers allowed me to post flyers for the green parenting study. Many local parents also allowed me to post virtual flyers on green parenting Facebook groups.

I also have scores of interview participants to thank. None of this research would have been possible without their selfless and generous gifts of time. I enjoyed every second of our moments together.

Finally, I would like to thank Courtney Morales, Lexington Books Acquisitions Editor, for encouraging me to stick with the book through all of the COVID-19 delays. In addition, Sam Brawand, of Brawand Consulting, provided outstanding last-minute editing to help me complete the manuscript. I so appreciate the team effort.

Some of the material in this book appeared in three of my previously published articles including: "Parenting for Environmental Change," *Contexts* (Winter 2022): 26–31; "'Before She Was Born I Ate Cheerios and Beer for Dinner': A Qualitative Examination of Green Parenting in Lowcountry South Carolina," *Humanity & Society* 45 (4) (2021): 1–32; and "'I'm a Normal Person': An Examination of How Utility Cyclists in Charleston South Carolina Use an Insider/Outsider Framework to Make Sense of Risks," *Urban Studies* 48 (7) (May 2010): 1439–56.

Introduction

My initiation into green living began when I was an undergraduate at Virginia Tech. While living in Blacksburg, I discovered the Appalachian Mountains and spent many weekends traversing hiking trails and exploring swimming holes. About my third year in college, a friend, who was leaving for the Peace Corps, gave me their Cannondale bike. This was a game changer. I hold many fond memories of biking to classes, to my bartending gig, and to friends' houses. I was light and free. I moved often after those years and bike commuted whenever the local infrastructure allowed. In 2011, I landed in South Carolina for a faculty position at the College of Charleston. Today, I live in the suburbs across the Ashley River from the College. Except for the bravest of cyclists, the bridges over the Ashley River are too dangerous for biking. The College is only a few miles from my house, yet I drive to work. When Gaspard, our child, became old enough to ride a bike, we were thrilled to start biking together to a few local amenities. Lately, the neighborhood basketball courts are our favorite destinations. My own love of utility cycling and my frustrations with the Lowcountry infrastructure served as the motivation behind my original bike commuting research.

More recently, I began research on green parenting. Our household started, stopped, and continued many sustainable living activities. We ditched a backyard composting project due to rat infestation but successfully contracted with a local company to collect and compost our waste. I remember the delightful day that Gaspard's learning center conducted a ladybug release for natural pest control. The children giggled as they walked around the garden and freed the tiny polka-dotted beetles onto leaves and branches. We joyfully replicate this ritual in our own yard twice a year. At the same time, we struggle to keep single-use plastics out of our dual working-parent home. These are the family experiences that inspired me to think seriously about both the simultaneous contradictions and rewards of green parenting.

METHODS

This book on sustainable lifestyles captures the experiences of everyday people who identify as "green parents" or as "bike commuters." I discuss forty-eight in-depth qualitative interviews conducted with green parents in 2016 and, to a lesser extent, two earlier sets of interviews conducted with bike commuters.[1] While this book focuses primarily on the green parenting study, I noticed parallels in the findings from all three studies. Where relevant, I discuss the two sets of cyclist data. One set of data represents choice utility cyclists (those who can afford a car but bike by choice) and the other represents necessity utility cyclists (those who cannot afford a car and bike by necessity). The two studies on biking define "commuting" as any utilitarian use of the bike. The book employs both the terms "utility" cycling and bike "commuting" to refer to the same activities.[2] When quoting interviewees, the green parenting study is abbreviated as "GP," the choice bike commuting study as "CB," and the necessity bike commuting study as "NB."

Respondents varied quite a bit in terms of the variety and intensity of their green activities. For instance, parents, who self-identified as "living green," named a wide range of activities (cloth diapering, recycling, composting, eating locally sourced produce, minimizing purchases, repurposing clothing and toys, etc.) that they considered sustainable. Like the green parents, bikers, who self-identified as "commuters," also referred to a wide range of activities including cycling to the store, to work, to run errands, to friends, and to recreation destinations.

WHAT WE KNOW ABOUT SUSTAINABLE LIVING

Approximately 94% of US citizens express a general concern for the environment and 77% express a desire to make more efforts to live sustainable lifestyles.[3] This is despite the country's enormous and negative impact on the global environment. The United States, one of the top ten countries in terms of its ecological footprint (Global Footprint Network 2017), is also the second-largest contributor to global climate change (Crippa et al. 2020). What explains this vast gap between the lack of national political determination to contribute to environmental sustainability and the massive numbers of people who express a desire to live green lifestyles? One important answer is that green living is a deeply symbolic set of social practices and social innovations that are enacted daily on a highly contested political terrain.

This book is about how and why regular people, who live in the politically conservative Lowcountry South Carolina, make the effort to navigate cultural and other challenges to carry out green lifestyles. By examining the motivational, regional, and sociopolitical underpinnings of people's understandings of green living in the Lowcountry, I reveal a diverse, yet connected, continuum of daily pro-environmental practices. The practices sometimes appear distinct, such as food gardening versus bike commuting, but the practitioners often share common ideals (like wanting to feel more social connection). I show that environmental practitioners draw on a holistic set of meanings and motivations in their understanding of green living. In addition, as I will discuss in chapter 4, these findings echo other studies which document that, while many people want to reduce their ecological footprints, their conceptions of green living are complex and extend to the wider desires of connecting to people and nature, living a meaningful life, and experiencing joy, among others.

The interdisciplinary field of environmental sociology is an ideal discipline from which to carry out research on sustainable lifestyles. Work generated out of this subfield of sociology focuses on green consumption (Frezza et al. 2019; Schmidt et al. 2014; Schor and Thompson 2014; Tullia 2013; Willis and Schor 2012), local food movements (Barron 2016; Carolan 2018; O'Neill et al. 2019; Robinson 2016), ecovillages (Chitewere 2018; Dias et al. 2017; Liftin 2016; Ulug et al. 2021), social innovation (Backhaus and Genus, 2017a, 2017b; Bergman et al. 2010; Cohen et al. 2013; Edwards-Schachter et al. 2012; Haxeltine et al. 2017; Jaeger-Erben and Ruckart-John 2015; Lestar and Bohm 2020; Mert-Cakal and Miele 2021; Moulaert et al. 2013; Olsson et al. 2017; Spijker et al. 2020), and social practices (Burk 2017; Godin et al. 2020; Halkier et al. 2011; Hards 2011; Hargreaves 2011, 2016; Hasselqvist and Hesselgren 2019; Huddart et al. 2015; Kennedy et al. 2015; Lorenzen 2012, 2014a, 2014b, Meyer and Kersten 2016; Shove et al. 2012; Shwom and Lorenzen 2012; Speck and Hasselkuss 2015; Spijker et al. 2020; Spotswood et al. 2015; Strengers and Maller 2015), among others. One comprehensive edited volume on green living that touches on a number of the above topics and theoretical frameworks is by Meyer and Kersten (2016).

The Greening of Everyday Life: Challenging Practices, Imagining Possibilities

The subfield has made less progress, however, on the topic of green parenting. Notable exceptions include an exploration of the role that children play in connecting parents to each other to build green social capital (Woods et al. 2013), a theoretical examination of how to think about family household dynamics in research on responses to climate change (Gibson et al. 2010) and analyses of the question of whether becoming a new parent serves as a possible "moment of change" for the development of sustainable consumption habits (Burningham and Venn 2020; Burningham et al. 2014). In a recent longitudinal study, Thomas and colleagues (2018) examined the impact of parenthood on environmental values and behavior. One area of green parenting that has generated significant analyses is research on how individualized solutions, at the household level, often exacerbate a culture of "intensive mothering" (Afflerback et al. 2013; Braun et al. 2020; Cairns et al. 2013; Cairns and Johnston 2018; Cousins 2021; Huddart et al. 2019; MacKendrick 2014; MacKendrick and Cairns 2019; Parker and Morrow 2017). Studies also document how environmental risk is individualized for mothers who must find ways to adjust their mothering and pregnancies to a wide array of factors outside of their control (Crighton et al. 2013; MacKendrick and Cairns 2019). A few exceptions to this focus on "intensive mothering" do exist. Kelly Dombroski (2016), for instance, concludes that mothers' everyday labor, especially the work of going diaper-free (the author uses the term "nappy-free"), is a type of environmental activism inside the home.

More progress has been made in terms of researching bike commuter lifestyles (Burk 2017; Fincham 2006, 2007; Hasselqvist and Hesselgren 2019; Hendlin 2016; Horton et al. 2007; Jones 2005; Kidder 2005, 2006a, 2006b, 2008, 2011; Larsen 2018; Lee 2015; Skinner and Rosen 2007; Spotswood et al. 2015). Elizabeth Wilhoit and Lorraine Kisselburgh (2017), in their examination of bike commuting in the American Midwest, show that, while commuting is a form of resistance, cyclists often do not intend to resist through the act of biking to work. They argue that bike commuting serves as a hybrid form of activism that is not limited to human agency. Focusing on equity issues, Melody Hoffmann (2016) documents how bike advocacy has largely ignored the interests of economically disadvantaged groups and people of color. The final chapter of her book looks at ways that the cycling lifestyle, if made more representative of the interests of diverse populations, could serve as a potential site for social change. More recently, Patrick Rerat (2021) explored the rise of the e-bike. In his article, he shows that, while the e-bike cyclist shares many motivations with the conventional bike commuter, the e-bike technology has the potential to empower more diverse groups of people.

Popular press articles, books, websites, zines, blogs, etc. have paid quite a bit of attention to the topics of green parenting and bike commuting. There are several books that are marketed to the children of green parents (or, more accurately, to parents that are looking for "green" themed story books), for instance, *That Fish on Your Dish* (Marcus and Marcus 2013). Several memoir-style (Montgomery-Fate 2012) and journalistic (Louv 2008) books have been published. *Cabin Fever: A Suburban Father's Search for the Wild* (Montgomery-Fate 2012) is typical of an emphasis in popular literature on the importance getting children "outside" and "connecting children to nature."

As I discuss in chapter 4, this was also a strong theme for participants in the green parenting study. *The Eco-nomical Baby Guide* is a popular guidebook that was mentioned by interviewees in the green parenting study (Kelley et al. 2010). A more recently published guide is *The Sustainable(ish) Guide to Green Parenting: Guilt-Free Eco-Ideas for Raising Your Kids* (Gale 2020). While this book emphasizes "guilt-free" ideas, it does so in a media environment that tends to promote individualized responsibility. Take, for instance, this Facebook page (https://www.facebook.com/SustainableParentingAssoc), which currently features a photograph of cloth diapers hanging on a clothesline (an extremely labor-intensive activity) and hosts articles on even more laborious activities, such as an interview with a man who boasts that he grows and forages for 100% of his food. One notable exception to this trend is Shannon Brescher Shea's (2020) book, *Growing Sustainable Together: Practical Resources for Raising Kind, Engaged, Resilient Children*, which provides parenting advice, and a holistic view on what it means to live sustainably. Rather than define green parenting in terms of cloth diapering and household energy efficiency, Shea's message is that becoming closer as a family and community is part of becoming closer to nature. Her message is echoed throughout the narratives of the participants in the green parenting (GP) study.

In terms of bike commuting, witness the online magazine *Momentum* and the gender- focused zine *Taking the Lane*. "How to" bike books have been popular for a couple of decades and point to what some researchers call a biking *renaissance* in the United States (Pucher et al. 2011). *The Bohemian Guide to Urban Cycling* is one such guide (Benesh 2014). Several recent popular books describe the resurgence of bike commuting and advocate for bike-friendly urban planning (Colville-Andersen 2018; Reid 2017). Interest in bike commuting is reflected, as well, in participation rates. From 2000 to 2019, for instance, the number of US commuters who bike to work increased by 61% (Bike Advisor 2019).[4] Average bike commuter travel times have also increased approximately 10% (25 minutes to 27.6 minutes) over the 14 years between 2006 and 2019 (Burd et al. 2021).

The abundance of publications about green parenting and bike commuting in popular media outlets points to an increased public interest in sustainable lifestyles as viable responses to a host of environmental problems. This growing public interest, combined with our worsening environmental dilemma, suggests there is a need for social scientists to increase the production of research on green lifestyles. This book describes two green lifestyles and contributes to an empirical understanding of the pathways to, challenges of, and meanings/motivations of green living. The book also outlines the limits and potentials of some of the interventions, especially green media, that promote green lifestyle change.

As I cited earlier in this chapter, social practice and social innovation theories provide useful lenses for examining green lifestyles. These two perspectives are woven throughout the book, and are emphasized in chapter 4, where I discuss the motivations and meanings of green living. The two frameworks complement each other in how they emphasize different points in the trajectory of green lifestyle change. Where the social innovation perspective focuses on the introduction of agentic intentional change, the social practice perspective focuses on the ongoing mechanics of how change emerges, is maintained, and decays over time in the habits of people's daily lives (Hochgerner 2011). In chapter 3, when discussing parents' challenges with media consumption, I draw on the attention economy literature to shed light on how and why parents experience green media as both a resource for and a hindrance to sustainable living. Throughout the book, where useful, I view the data through additional theoretical lenses.

WHY STUDY GREEN LIFESTYLE CHANGE?

As Paul Crutzen and Eugene Stoermer (2000) first articulated, human activity has become the dominant influence on the earth's environment (conceptualized as a new geological age they coined "the Anthropocene"). In the Anthropocene, human activity has caused disruptions of plant and animal life cycles (UNEP, United Nations Environment Programme 2010, 2021a); resource depletion (Jackson et al. 2001); declines in biodiversity (Loreau et al. 2022); depletion of the ozone layer (Ripple et al. 2017); and increased deforestation (Austin et al. 2017), wildfires (UNEP 2021a), and extinctions (Urban 2015); rising pollution levels (UNEP 2021a); and global warming (Intergovernmental Panel on Climate Change, IPCC 2022), among many other interlocking and devastating problems.

Individuals and communities do not suffer equally from the threats of environmental devastation nor do they share equally the culpability of contributing to it. Just as individuals of low socioeconomic status (Hajat et al. 2015), people of color (D. Taylor 2014), women (Buckingham and Kulcur 2009), and indigenous communities (Fernandez-Llamazares et al. 2020), face disproportionate threats from environmental deterioration, upper-income earners and nations exert an unequal and higher negative impact on the environment (Weidmann 2020). For instance, the world's top 10%, in terms of income, are estimated to be responsible for 25% to 43% of humanity's negative impact on the environment. By contrast, the world's lowest 10%, in terms of income, are estimated to be responsible for 3% to 5% of humanity's negative impact on the environment (Teixido-Figueras et al. 2016). In other words, wealth and inequality drive a significant (Chancel and Piketty 2015) portion of anthropocentric environmental degradation. In a summary of the literature on affluence, consumption, and ecological deterioration, Weidmann and colleagues (2020) found that three key factors explain why and how highly affluent consumers disproportionately and negatively impact resource use and depletion. First, they consume at high levels; second, they have an interest (through employment, ownership, investments, etc.) in supporting the growth paradigm upon which our current global, capitalist, economy is based; and third, they drive the consumption norms of the local, regional, and national places in which they live (think of how luxury consumption represents an important marker of social status). Furthermore, they point out that, in many areas of the world, low- income groups have moved into middle- and upper-income lifestyles. This will further exacerbate the ability of technological, institutional, corporate, and governmental programs to achieve sustainability. As such, in addition to systemic-wide solutions, for example the transformation of our consumption and growth-based global economy, it is equally important to promote green lifestyles. In making this argument, the authors point out that this does not mean denying low-income earners, or anyone else, the chance to meet basic needs. Well-being, they note, has been shown to increase with rising income only up to a point and then it diminishes as income and consumption rise further. They conclude that people living and modeling sustainable lifestyles have the potential to influence the economy and consumer culture in a green direction, from the bottom up (Weidmann et al. 2020).

A large body of diverse interdisciplinary work, both applied and empirical, theorizes and documents how substantive environmental change is indeed possible at the level of daily life (Haswell 2021). Researchers who write about the "behavioral wedge" (Dietz et al. 2009; Macias and Williams 2016; Vandenbergh et al. 2011), for instance, have documented sizeable reductions in energy waste when consumers adopt new habits and technologies at the household level. More recently, the Center for Behavior and the Environment produced a comprehensive report, *Climate Change Needs Behavior Change*, that documents and ranks thirty effective behavioral solutions to climate change. They conclude that "high impact pro-environmental behavior, like reducing food waste, is beneficial not only in terms of their mitigation potential but also in terms of economics, human health, and well-being" (Williamson et al. 2018, 16). Most of this work emphasizes that, due to the Jevons Paradox, the most important dimension of green consumption is reduced consumption because sustainable purchases, especially those that lead to cost savings, can encourage practitioners to consume at higher levels (Alexander and Ussher 2012). Given that individual households are estimated to contribute as much as 60% of all global greenhouse gas emissions and almost 60% of all global land, material, and water use (Ivanova et al. 2016), green consumption is not just possible but also, arguably, imperative for achieving both social and environmental justice (Kalbar et al. 2018; Matusik and Koci 2019; UNEP 2010).

This book's assumption that daily life is an important source of social change lies in contrast to views expressed by other camps of thought (Shove 2010). An important body of empirical and theoretical work, for instance, critiques that green consumption and lifestyle programs displace attention away from collective action (Lorenzen 2014b; Maniates 2002; Szasz 2007; Werfel 2017). Many document, further, that green consumption shifts environmental blame from the industries and governments that caused the problems to individual citizens (Kennedy et al. 2015; MacKendrick 2010, 2018; MacKendrick and Stevens 2016; Washburn 2014). Other studies have pointed out that increased green consumption is still excessive consumption (Stuart et al. 2020). Some studies find that consumers command very little control over the larger supply chain of goods and services (Fuchs et al. 2016), and that, therefore, green consumption approaches simply do not result in net gains of pro-environmental behavior (Mazar and Zhong 2010; Olson 2012; Tiefenbeck et al. 2013). Mazar and Zhong (2010) criticize that the good feelings that come with practicing green consumption, like updating insulation in one's home, can negatively "spill over" as practitioners use the positive green activity to justify other behaviors, such as keeping the heat on high. It has been argued, as well, that policies based on household consumption can *nudge out* support for more substantive institutional- and governmental-level policy and infrastructural changes (Hagmann et al. 2019; Stuart et al. 2020; Werfel 2017). It is important to note, however, that the David Hagman and colleagues (2019) study found that *nudge out* disappears when practitioners are provided with information about how such approaches do not represent sufficient replacements for systemic level change. These represent just a few of the many critiques of sustainability interventions which focus on lifestyle changes.

This book concurs with the substantial empirical evidence that green consumption, and other policies that focus exclusively on green lifestyles, face limits; they can be ineffective and displace responsibility away from systemic political and economic forces, among other challenges. However, I assert that green lifestyles, though not sufficient in-and-of-themselves, represent one of many necessary paths to sustainability. As Haswell (2021, 30) suggests in his essay on why individuals should consume less even if wider society does not: "Peer-to-peer influence, cultural norms, base instincts, education, knowledge, critical thinking, and individual reflections are also undoubtedly integral to lifestyle choices. Externalities and macro-level factors can present strong drivers of behavior, but we must not entirely absolve ourselves of responsibility at the individual level." Black (2020) also makes the important point that green living and consumption are sometimes the only accessible means available to people concerned about the environment. I found this to be especially true of time-strapped individuals, like Lowcountry green parents. Keeping the thermostat low in the winter might involve a little discomfort but involving oneself in environmental advocacy in the public sphere entails potentially large investments of time and of energy. Finally, Sara O'Shaughnessy and Emily Kennedy (2010, 553) make the important argument that the traditional view of activism as a set of activities that takes place only in the public sphere ignores the "myriad efforts behind the scenes that also constitute environmental activism." They go on to clarify that women, especially, have a history of engaging in relational activism "that utilizes relationships among networks of like-minded individuals and blurs the distinction between public and private spheres by using daily behaviors as the locus for social and environmental change."

Importantly, while green consumption is full of contradictions, green lifestyles include consumptive behaviors, but they also represent so much more. Numerous studies document that green consumption is part of a collection of sustainable identities, networks, lifestyles, and political action (Atkinson 2012; Dobernig and Stagl 2015; Evans and Abrahams 2009; Hamilton et al. 2018; Hedlund-de Witt 2017; Lorenzen 2014a; Micheletti and Stolle 2012; Noblet and Thogersen 2012; Scerri and Magee 2012; Schlosberg and Coles 2016; Stolle and Micheletti 2013; Willis and Schor 2012). In addition, solutions that emerge entirely from the top down will inevitably reflect the problematic worldviews that are embedded in current systemic political and economic forces. Finally, well-meaning infrastructure investments and planning (for example, installing bike lanes) have failed when they were not informed by how people actually carry out their daily lives (Hurtado 2018; Lubitow et al. 2016).

A LOOK AHEAD

Chapter 1: "Conservation Is What You Are Doing!: Living Green in Lowcountry South Carolina" maps out the regional context and describes the Lowcountry's sociopolitical culture and unique natural environment. I define *green lifestyles* and illustrate how the Lowcountry, home to a strong conservation culture and conservative politics, presents both constraints and opportunities to green living. I also describe the demographic characteristics of the participants in the three studies.

Chapter 2: "Pathways to Living Green" draws on several frameworks, including social practice theory, embodiment, and Siegel and colleagues' (2018) vision of pro-environmental behavior as a *living forest*, to provide context to the stories that practitioners tell about how they got started practicing green lifestyles. Most practitioners described how their lifestyles grew out of a mix of evolving experiences over long periods of time. Utility cyclists and green parents referred to an interwoven patchwork of origins including lifelong role modeling; exposure to education, media, travel, and the outdoors; important moments of change, like having a baby; and changes in resources and/or infrastructure, among others.

Chapter 3: "Green Parenting Challenges" focuses on the difficulties faced by parents in their efforts to enact green living. Some of these challenges emanate from the regional sociopolitical climate of the Lowcountry described in chapter 1. Here, I explore other challenges that are unique to the demands of green parenting, including time, money, and media. Bike commuters are not covered in this analysis as they primarily pointed to the cultural and infrastructural challenges that are already outlined in chapter 1.

I draw on the social practice, social innovation, and attention economy frameworks to discuss challenges parents faced and how they worked to overcome them. Parents overwhelmingly pointed to limited time and financial resources to make what they felt were good green choices. Parents also talked about connecting to each other and to nature at the same time they talked about wanting to disconnect from media. I show how this dual role of media explains the tensions that parents experienced between wanting to use media to enhance green living and, simultaneously, wanting to disconnect from media and reconnect with each other. In addition, while some parents sought green living guidance from media, others actively avoided it because, for them, it was unhelpful and/or confusing. While parents experienced significant challenges with time, money, and media, they also critiqued consumerist approaches to green behavior. In the following chapter 4, I explore how this critical attitude, along with their long-term focus on socializing their children into a greener next generation, enabled parents to successfully resist individualized guilt over what they were not able to accomplish.

In chapter 4, "Green Living Motivations and Meanings," I draw on social practice, social innovation, and embodiment frameworks to understand the forces that enabled parents and commuters to engage in green activities and develop green beliefs. I explore how parents and bike commuters imagined green living and how their stories connected to their everyday practices. Quite a bit of overlap exists in terms of how parents and bikers understood their sustainable lifestyles. In all three studies, participants emphasized enjoyment. Parents and choice commuters also shared a common focus on connecting to nature and people.

Green parents talked about how interlocking motivations (reducing environmental footprints, socializing children, ensuring families' health, and engaging in thrift) and meanings (conscious decision making, connecting to nature, connecting to each other, and cultivating a sense of well-being) directly informed their child-rearing practices. Through their parenting habits, they socially innovated to bring about multi-generational environmental change.

Commuters were asked about motivations but not meanings. Choice bikers named multiple interlocking motivations for commuting (exercise/recreation, environmental concerns, enjoyment, thrift, convenience, connecting to nature, connecting to each other, and self-sufficiency). Necessity bikers named some of the same interlocking motivations (enjoyment, exercise, and thrift). One necessity biker also pointed to convenience.

Finally, I discuss how most parents' stories focused on the way their pro-environmental beliefs drove their green behaviors. For cyclists, this was less true. Cyclists told more stories about commuting for personal gain, like exercise. However, many of the choice commuters also told how their current green awareness evolved out of the bodily act of utility biking.

In the conclusion, I explore the advocacy/policy implications of the results and pose questions for applied and interdisciplinary future social science research. I also briefly point to several Lowcountry nonprofit green initiatives that serve as examples for how to incorporate some of the programmatic/policy suggestions that emerged out of this book's analyses. I conclude that, while green living is full of contradictions that we cannot ignore, we need to take people's daily lives seriously. It is in our daily lives that we become aware of, digest, and ultimately impact larger societal ideas about environmentalism and what it means to live well.

GUIDING QUESTIONS

In the following pages, I ask several interlocking questions. What is green living in Lowcountry South Carolina and what challenges do practitioners face? What meanings, resources, and skills do practitioners draw on to overcome challenges? What are the pathways to green living for parents and bike commuters? What motivations drive practitioners and what meanings do they draw from green living? How do major life transitions, like becoming a parent, impact our daily green habits? How can we leverage those life transitions to enact green change? How can social practice, social innovation, *living forest*, embodiment, and attention economy frameworks help us to understand the experiences of green parents and bike commuters? How can these results inform sustainable policies and interventions?

When I began studying environmental issues in the 1990s, I started by asking abstract questions about "other people out there." With my research on green living, I came full circle back to my own visceral joy as I bike down the block to get an ice cream cone or hit the basketball courts with my child. As Michael Carolan (2009, 5) eloquently observes, "We can only know nature as bodies-in-the-world: as bodies *doing* nature, from somewhere, in a particular way." The narratives that unfold throughout the rest of this book tell stories about reconnecting to each other and to nature and the bodily pleasures of, say, setting out a blanket and enjoying a pretty day reading a book with your child in the park (Kareena, 2 children, GP, I-29).[5] These may not seem like radical stories, but in their telling, everyday people are rewriting their relationships to each other and to nature in more compassionate and relevant terms. Policy makers should listen.

NOTES

1. The interviews with green parents and choice cyclists lasted from forty-five minutes to an hour and a half long. Both of the bike commuting studies covered the same broad research questions, how people got started and what they liked and did not like about commuting; however, the interviews with necessity cyclists were only a few minutes long. Choice cyclists (CB) and green parents (GP) were very motivated to discuss their lifestyles and eagerly responded to my flyers to meet with me at my office, their homes, playgrounds, or coffee shops to talk about their activities and motivations. Necessity cyclists (NB) did not respond to my flyers. Instead, I recruited them by approaching people, who had visibly arrived by bike, at public laundromats, libraries, shelters, soup kitchens, and public venues, where they graciously offered a few minutes of their time. The data from the necessity commuting study is, therefore, rich with information but light on detail.

For all three studies, I conducted face-to-face, open-ended interviews. Student research assistants participated in the collection of interviews with necessity cyclists. In addition, for each study, I used semi-structured interview guides (appendices A, B, C), along the lines of naturalistic inquiry (Lincoln and Guba 1985), to facilitate a conversational dialogue and develop a holistic understanding of individuals' perceptions within a particular social context (Erlandson et al. 1993).

I used NVivo software to carry out inductive qualitative thematic analyses in which I began with open coding on the interview data to generate themes and subthemes. Then, I categorized themes into a codebook. Finally, I used the codebook to produce focused coding. This coding method allowed my theoretical frame to evolve as I iteratively went back and forth between reading the literature and coding the data. This standard coding method is explained in classic texts such as Robert S. Weiss's (1994) *Learning from Strangers: The Art and Method of Qualitative Interview Studies*.

For the green parenting study, occasionally a spouse would come with the primary interviewee and participate. When presenting the interview data, I make it clear when a secondary interviewee adds comments to an interview. For the green parenting (GP) and choice commuter (CB) studies, respondents provided their own pseudonyms. For the necessity commuter (NB) study, my student research assistants made up the pseudonyms. One interviewee from the green parenting (GP) study wished to be quoted with her real name—Germaine. Germaine Jenkins runs an urban farm in North Charleston, started a grocery store in a food desert, and is a regionally well-known and strong advocate for sustainability issues and for exposing and eradicating environmental racism. When citing quotes from the three studies, I provide the pseudonyms and interview numbers. For the parents' quotes, I also provide the number of children in their household.

2. Since less than 20% of daily travel (U.S. DOT, Department of Transportation 2015), by any mode, is estimated to be work-related, I studied the broad category of utility cycling rather than the narrow category of work-related bike commuting.

3. This study was made up of 79% Americans and 29% Australians (Southern Cross University 2019).

4. Among all US citizens who utility cycle, younger workers are increasingly the most likely to bike to work (Burrows, 2019).

5. When quoting green parenting interviews, I provide the pseudonym and the number of children in the household. The initials "GP" indicates that I am talking about green parenting data. When referring to commuter interviews, I provide the pseudonym. The initials "CB" indicates that I am talking about choice biker data and the initials "NB" indicates that I am talking about necessity biker data. For all three interview studies, I numbered the interviews. For instance, for the green parenting study, I conducted forty-eight interviews, and this is a quote provided by interview #29.

Chapter 1

Conservation Is What You Are Doing!

Living Green in Lowcountry Charleston

My first experience with the Lowcountry occurred long before I ever moved to the region. It was in the reading of Pat Conroy's (1986, 5) fictional book: *The Prince of Tides*. Conroy's character explains: "To describe growing up in the Lowcountry of South Carolina, I would have to take you to the marsh on a spring day, flush the great blue heron from its silent occupation, scatter marsh hens as we sink to our knees in mud, open an oyster with a pocketknife and feed it to you from the shell and say, 'There. That taste. That's the taste of my childhood.'" Conroy's words came back to me as I made my first ever drive, in the summer of 2001, to Folly Beach—windows open, as I crossed over marshes and inhaled the rich smell of fresh *pluff* mud. Over the next twenty-plus years, I came to love and admire this strange and beautiful land of oppressive heat, never-ending gnats, and other attacks on the senses. The Lowcountry offers tough love. It was this *country*, and motherhood, that explains why I now know how to string a fishing rod, hook and all. It is why I also know that the secret to successful crabbing is a half-open can of cat food. I love this place where strangers feel comfortable calling me "honey" and yet I dream, often, of escaping to a cooler climate and where I'd also see fewer trucks with modified smokestacks "rolling coal."[1] Green parents and bike commuters navigate this social and geographical space, which includes families and friends, as they enact their green living. Before describing the conservationist and conservative cultural backdrop of the Lowcountry in more detail, I will first provide information about the region's geography and the demographic characteristics of the three studies.

LOWCOUNTRY SOUTH CAROLINA GEOGRAPHY

The Lowcountry, as the name suggests, is *low*, at about an average of 4 meters, or 13 feet above sea level. Lowcountry South Carolina sprawls across the southern part of the state and extends from the coast to the fall line (a geographical boundary that demarcates the division between an upland and lowland region). Charleston, the second-largest and oldest city in South Carolina (World Population Review 2022), is a medium-sized coastal metropolis located in an Atlantic inlet formed by the Ashley, Cooper, and Wando rivers. Like many regions in the southern/coastal United States, the city of Charleston, with a population of 150,227 in 2020 (compared to 120,083 in 2010) (U.S. Census Bureau 2022a), is one of the fastest growing in the United States (US News and World Report 2022). In 2021, Charleston County, which is also rapidly growing, had a population of 413,024 (compared to 350,209 in 2010) (U.S. Census Bureau 2022a).

DEMOGRAPHIC COMPARISONS OF THE GREEN PARENTS AND UTILITY CYCLISTS

The demographic makeup of the three studies both reflects and diverges from Lowcountry and national population characteristics. The incomes and education levels of Charlestonians are somewhat higher than the national norms.[2] For the green parenting (GP) study, White respondents were overrepresented and Black respondents were underrepresented, while Hispanics were somewhat representative of regional demographics. The green parenting participants were also more educated than regional averages. For the forty-eight interviews with green parents, participants were overwhelmingly White (90% White, 6% Hispanic, 4% Black), female (96%), well educated (100% completed high school and 79% completed a bachelor's degree or higher), employed (77% employed and 23% homemaker with employed partner), and married (93%). Eighty-one percent of the parents were aged *thirty-something*. While this sample is Whiter, more educated, and more female than Charleston's population, it does reflect findings that *mainstream environmental ideals*, as opposed to *pro-environmental behavior* and *environmental justice ideals*, are more likely to be held by White and well-educated segments of the population (Gifford and Nilsson 2014).[3] It also reflects findings that women report higher levels of concern about environmental issues than men (Carfagna et al. 2014; Elliott 2013; Gifford and Nilsson 2014). Some evidence suggests that this gender divide is especially true for White women and men.[4] Eighty-five percent of the green parent households had children aged four years or younger.[5] Most families had one or two children at the time of the interviewing. Fifty-four percent had one child, 35% two, and 10% three or more. This reflects national trends in which US mothers, especially those who are well-educated, tend to have two children by the end of their childbearing years.[6]

For the choice cycling study, men and Whites were overrepresented. The members of this study were also more educated than regional averages. Of the forty-one interviews with choice cyclists (CB), the participants were overwhelmingly White (98%) (one Black woman participated in this study), male (66%), and well educated (100% completed high school, 83% completed a bachelor's degree or higher, and 0.05% were currently working on bachelor's degrees). In terms of employment, 89% of the choice cyclists were employed and all of them either owned their own businesses, worked in professional positions, or worked in the arts. The remaining cyclists were college students and retirees.

For the necessity cyclists (NB), men and Blacks were overrepresented. Participants were also less educated than regional averages. Of the twenty-one people interviewed, the participants were largely Black (68%), male (89%), undereducated (57% completed high school, 14% either completed the GED or were working on it, and 10% completed either an associate's or a bachelor's degree), and underemployed (48% were unemployed of which half were on disability, 10% were retired, and 43% were employed in blue-collar, low-paid service, and/or informal economy work).

Age categories for choice (CB) and necessity commuters (NB) were recorded differently than for parents. However, choice commuters tended to be both younger and older than the green parents (GP) (17% of choice commuters were under the age of 26 years, 49% were within the age range of 26 to 45 years, and 34% were over the age of 45 years). Necessity commuters tended to be older than both the green parents and the choice cyclists (42% were over the age of 49 years).[7]

The choice cyclists had lower marriage rates than the green parents. Sixty-three percent were married or cohabitating with a life partner. The necessity cyclists had the lowest marriage rates with only 0.3% indicating that they were either engaged or married. The middle age and high marriage rates of participants in the green parenting study can be explained by the fact that these participants were in the life stage of raising children. The overrepresentation of men, in both cycling studies, is reflective of the gendered characteristics of bike commuting in the United States. According to the U.S. Census Bureau (2016), though the numbers of women who report bike commuting have increased, the numbers of men who bike commute still exceed that of women. Men made up 77% of bike commuters in 2006 and 72% in 2014. Women made up 23% of bike commuters in 2006 and 28% in 2014.

To more fully understand the parents and bike commuters represented in this set of studies, one must consider the uniquely conservationist and conservative culture, as well as the infrastructure challenges of the Lowcountry.

CONSERVATISM IN THE LOWCOUNTRY

It was only in the green parenting study that I directly asked parents about whether they considered themselves to be "environmentalists." Interviewees' attitudes toward environmentalism are reflective of the unique conservation culture of the region. They tended to define their green identities by what they do rather than their political orientation or any activist leanings. The following statement from Stella (1 child), the owner of a business that helps people set up food gardens in their yards, is typical of the sentiments of green parents.

> I don't really use the word "environmentalist." . . . I feel like it gets a negative connotation. [By] my actions and my career, people see that . . . I care about it. . . . It's so funny to me, 'cause my in-laws are very conservative politically, but they're all environmentalists . . . land conservationists. My father-in-law's a crazy big hunter. He manages thousands of acres for trees, which is a renewable resource. It's funny when they're like, "those environmentalists," and I'm like, "but you are too. Conservation is what are you doing!" (I-37, GP)

Likewise, Sharice (2 children), who grew up with parents that canned food, gardened, and drove recyclables to a recycling unit, recalled:

> I don't remember my parents saying "this is better for the environment" 'cause we didn't use the word "environment." My dad did not use pesticides [but] he laughs at the organic stuff and says "they don't realize what organic really means. Even though I was growing my own yard, trying to avoid pesticides, it still wasn't completely organic." (I-18 GP)

Another mom, Ray (1 child), noted that she reuses tinfoil, and washes and dries it by hanging it on a repurposed bottle-drying rack. She talked about how: "It actually surprises some of our friends, to find out how much we reduce, reuse, and recycle. We don't trumpet it. I know some people are like, 'Oh yeah, we recycle everything.' We just do it 'cause it's something we always did" (I-37, GP).

In sum, when asked, "Would you call yourself an environmentalist?" only fifteen of the forty-eight green parents answered that they accepted that title. Twenty parents indicated that they identified with the "environmentalist" title but felt they either *cannot* or *do not* live up to it all the way. Nine parents resisted the title because of its negative connotation. Nine of the parents also brought up, without prompting, that they felt that we need to get "beyond labels."

I did not ask bike commuters if they consider themselves to be "environmentalists." However, I did ask about their motivations for bike commuting and their answers indicated that they likely do not identify very strongly with an *environmentalist* label. Only thirteen of the forty-eight choice utility cyclists named environmental concerns as one of their original motivations for bike commuting, though thirty indicated that, at the time of the study, they were interested, or newly interested, in environmental issues. Only one of the twenty-one necessity cyclists indicated that environmental concerns represented an important motivation for starting bike commuting. None of the necessity cyclists brought up environmental issues when talking about current motivations. The tepid interest in environmentalism among cyclists and the resistance to the environmentalist label among many parents in these studies may reflect how environmental views have become more partisan over recent decades (Anderson 2017). It also reflects the unique culture of the Lowcountry region.

CONSERVATION IN THE LOWCOUNTRY

The Lowcountry subtropical coastal region is home to incredible opportunities for connecting to nature. This verdant-humid-flat tangle of marshes, creeks, rivers, islands, public beaches, and several sprawling parks is teaming with wildlife, including everything from iconic bottlenose dolphins, white ibises, and alligators to the more common green anole lizards, oysters, and crabs. If you have got a simple piece of string and a couple of leftover chicken bones, then a full day of crabbing and other adventures lie ahead of you and your children.

Lowcountry residents work hard to preserve the environmental amenities and resources of the region. These conservation efforts, as we know them today, began in the wake of the 1989 Hurricane Hugo, which damaged 80% of the Charleston peninsula's downtown homes and caused widespread devastation across the surrounding areas. Out of this ruin and rebuilding, locals drew on the region's traditional emphasis on property and hunting rights (Halfacre 2012) to marshal a strong conservation culture. While this conservationism keeps strengthening, it occurs amid the historical context of the politically conservative South. With the exception of Jimmy Carter, South Carolina voted Republican in every presidential election since 1964. Like most states with low per-capita incomes and Republican voting patterns, South Carolina is least likely to support strict environmental regulations (Bialik 2016). A moderate oasis in a "red" state, Charleston has elected, for well over one hundred years, a Democrat into the mayoral office. Its current mayor, though an Independent, was endorsed by former longtime mayor Riley, a Democrat.

More recently, Lowcountry demographics evolved as workers, attracted by growing IT and industrial jobs (Moore 2018), migrated to the area. These evolving demographics are reflected in the interviews with green parents, as over 50% of the respondents reported that their global and cultural exposure influenced their green behaviors. For choice utility bikers, thirty-nine of the forty-one interviewees talked about how they had lived or traveled in areas outside of Charleston that were much better suited to bike commuting. The necessity cyclists did not talk about travel to other regions.

The immigration of people from other parts of the country and the world is both enabling and disabling the practice of greener lifestyles in the Lowcountry. People who have experienced green and bikeable communities elsewhere are bringing their ideas and lifestyles to Charleston. Several parents noted, for instance, that after they moved to the region, they started recycling and other green programs in their workplaces. One mother, Seneca (2 children), moved to Charleston from Dayton, Ohio, to work in the restaurant industry. At her first job, she started a wine cork recycling program that is still in operation today. Now she owns her own boutique selling secondhand children's clothing and toys. She refers to her suburban brick-rancher home as "the farm" where her family keeps chickens, a rescue dog, and one rescue pot-bellied pig (other piglets froze to death and the sow rejected the two that survived) (I-25, GP). Likewise, many choice bike commuters brought their cycling habits to Charleston. Luca explained that he "grew up in Africa where many people longed to have bicycles and very few people had cars. So, the whole notion of getting along without a car wasn't foreign to me" (I-39, CB).

LOWCOUNTRY INFRASTRUCTURE CHALLENGES

Newcomers to the Lowcountry brought green habits but they also contributed to rising population levels. The city of Charleston was ranked by the *US News and World Report* (2022) survey as the nineteenth-fastest-growing city in the country. As one document describes, a booming manufacturing and IT economy has transformed the Lowcountry from a small-scale tourist destination into a sprawling coastal metropolis (Huntsman and Steinberg 2019). This has led to the loss of open space, increased development, and traffic congestion, among other problems (Slade 2018). The growing sprawl, and associated increases in housing costs, forces residents to live further from the urban core, increases their commute distances, and makes activities, such as biking, more difficult. Some of the region's outer suburbs, also, offer minimal sustainable living amenities, including household recycling. The Town of Summerville, for instance, only recently restarted its curbside recycling program after shutting it down (for anything other than cardboard or paper) for five years (ABC News 2021). In 2016, North Charleston also shut down its curbside recycling programs in parts of the city (Rindge 2016). Those programs have not been restarted.

Increasing population levels put pressure on development and new development occurs on the only land that is left, ecologically fragile and flood-prone marshes. Developers regularly use fill (composed of crushed rock, sand, and clay) to raise new homes to the county and city's elevation requirements. The fill, however, does not absorb water as well as organic topsoil. This inevitably leads to increased flooding problems. In June of 2019, the city had to buy out forty flooded homes in a neighborhood in the West Ashley suburb of Charleston (Huntsman and Steinberg 2019). In spite of such building-related disasters, contractors continue to push forward with massive development plans in flood-prone areas (Donahue 2020; Liu 2020).

Participants in all three of the studies pointed to regional subpar infrastructure that is not conducive to green habits, like bike commuting. Though the region's car culture has historically been hostile to cyclists (Dille 2016; McCarthy 2011), it improved, in 2005, with the construction of the Wonders Way bike-pedestrian path on the Arthur Ravenel Bridge which connects the town of Mt. Pleasant to the Charleston peninsula over the Cooper River. Unfortunately, because travel on suburban roads in Mt. Pleasant is so dangerous, 73% of the people who use the Wonders Way path reported that they drive a car to access the path (McCarthy 2010). Future improvements are potentially underway as the city has approved \$18.1 million to retrofit the Ashley River Bridge with a bike-pedestrian path (Behre and Darlington 2018). As I write, it is now June 2022 and construction on the bike-pedestrian path has not yet begun, but we are all still hopeful.

The area's environment possesses physical assets that lend the space to year-round bicycle travel. The subtropical climate offers mild winters and warm summers (Charleston County Economic Development, CCED 2019); the flat terrain makes for easeful bicycling; and the natural and historic beauty provides for many scenic rides. Distances between destinations on the eight-square-mile peninsula, which makes up downtown Charleston, are easily traversed by bike, and the speed limits are fairly slow (an average of 30 miles per hour excluding the interstate) (Overcash 2009). However, most area residents live in the suburbs and cycling between peninsula Charleston and the outlying neighborhoods is, overall, dangerous (Behre and Darlington 2018; BCDCOG 2005).

As noted above, the Lowcountry possesses a number of physical assets that lend the city to year-round bicycle travel. However, Charleston County has the highest percentage of car crashes involving cyclists and pedestrians in all of South Carolina. Nearly 25% of bicycle crashes in South Carolina take place in Charleston County. South Carolina, itself, is one of the top ten most dangerous states for pedestrians and bicyclists in the United States. This is especially true for racially marginalized cyclists. While Black residents make up only 27% of South Carolina's population, they were involved in nearly 40% of bicycle crashes over a nine-year period (Palmetto Cycling Coalition and the South Carolina Livable Communities Alliance 2018).

Named by *Bicycling Magazine* as the "Worst City for Cyclists of 2016," Charleston's infrastructure and ordinances, like the policy of restricting bike parking to official bike racks, present challenges (Dille 2016). Cycling between the peninsula of Charleston over bridges to the suburbs and from one suburb to another is unsafe. One of the dangers posed to the area's residents is the lack of street connectivity in the suburbs outside of the Charleston peninsula (Kofsky and Bopp 2021). Often, to travel from one neighborhood to another, cyclists must traverse multi-lane, high-speed roads that provide no accommodations for bicycling (McCarthy 2011). Two cyclists were killed in just one year, 2020, on Savannah Highway (a four-lane divided highway that cuts through the suburb of West Ashley) (Powers 2021). On its north side, the peninsula connects to the city of North Charleston over land locally referred to as "the neck." The neck and North Charleston, like the other suburban areas, present cyclists with challenges in their attempts to travel between their neighborhoods, or between neighborhoods and the peninsula. The Cosgrove Bridge, which connects the Charleston suburb of West Ashley to the neighboring city of North Charleston, is one of the most dangerous treks. The six-lane, fifty-mile-an-hour bridge has no bike lanes, no break-out lanes, and no sidewalk. The two opposing sets of three lanes are divided by a concrete wedge that is only a few inches wide and on which plastic sticks, about waist high, are placed a few feet apart across the bridge. The wedge and sticks are intended to create a visual divide so that automobiles driving in opposite directions do not inadvertently crash into each other. Unfortunately, without any other way to walk or bike across the bridge, cyclists navigate this treacherous tightrope. It is truly terrifying to witness a cyclist walk a bike across this wedge, their bodies waving the sticks this way and that. Cyclists look for cars to speed by, and when there is a brief gap, they jump into the road and run past the stick, then hop back up on the wedge, and wait for the next opportunity. All the while, they must keep their bike close to their side, on the road, next to the cars whizzing by at fifty-plus miles an hour. A fifty-nine-year-old cyclist was struck and killed on the bridge two years ago (Hart 2020). This layout creates a scenario in which travel on and off the peninsula, and most travel in the suburbs, is best suited for automobiles (see fig. 1.1).

LOWCOUNTRY CLIMATE-RELATED CHALLENGES

Increasing sprawl and development are not just problematic for the pedestrian-bike infrastructure, they also exacerbate another serious challenge faced by the region; Charleston is experiencing dramatic impacts from global climate change (National Climate Assessment 2014). The region is vulnerable to major rain, flooding, and hurricanes. In 2015, it was the "100-year floods" where damage losses were estimated at $1.492 billion (U.S. Department of Commerce 2016). My own neighborhood was cordoned off. The water came up to the top steps, but did not make it into the house, and we had a picnic table floating in our backyard, but we had been able to elevate the outdoor air conditioning unit with extra bricks at the last minute. Some of our neighbors were not so lucky. Many had refrigerators and other appliances floating in their garages. Then came Hurricane Matthew in 2016, Irma in 2017, and Florence in 2018. Hurricane Irma also brought significant flooding (Darlington 2018). Both Matthew and Florence led to evacuations and Matthew caused major power outages, flooding, and wind damage (National Weather Service, NOAA 2016). For residents who were spared damage to their homes, the one good thing about these disasters was that the disruptions occurred over the course of a few days/weeks.

On an increasingly more regular basis, however, significant portions of the Lowcountry coastal region experience what is called "nuisance" or "sunny day" flooding. Between 2015 and 2017 alone, high-tide floods, including sunny day floods, increased from thirty-eight to fifty per year in the city of Charleston (Union of Concerned Scientists, UCS 2017). Through the 1980s, sunny day floods, which occur when tides pass the seven-foot mark, happened just five times a year and usually only when a hurricane pushed waves ashore. But in 2019, Charleston had a record eighty-nine days of sunny day flooding. As global sea levels rise, this is expected to get worse (Bartelme 2020). According to a recent report published by *Climate Change & Health*, South Carolina is one of the most vulnerable states in the United States to negative effects from climate change (McKillop et al. 2020).

Charleston is not alone in the dangers it faces from climate change and environmental degradation. A recent Intergovernmental Panel on Climate Change (IPCC) (2022) report calculates that as much as 45% of the world's population is highly vulnerable to climate change. Globally, one in three people are currently exposed to deadly heat stress and this is projected to increase to 76% by the end of the century. It is also projected that one in ten people will live in flood-prone coastal zones worldwide by 2050.

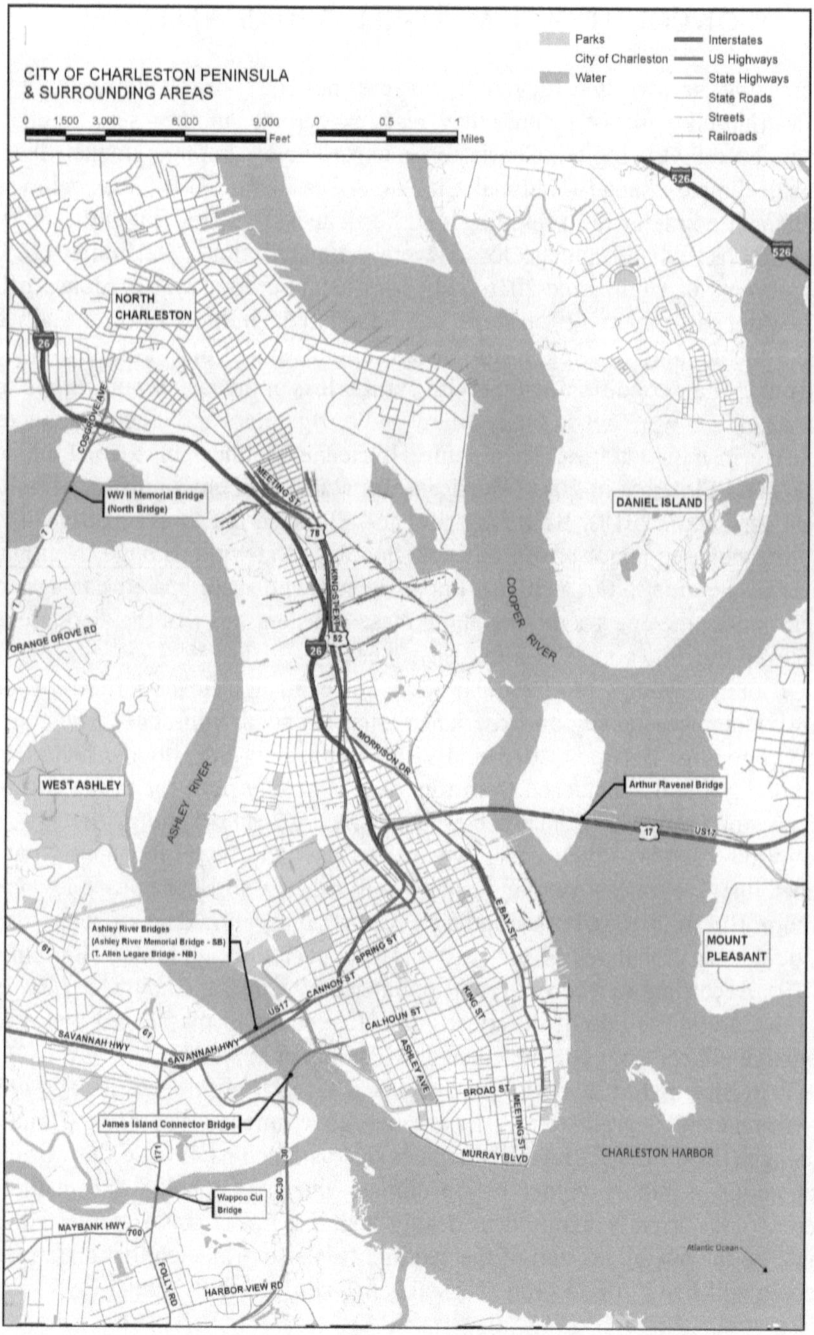

Figure 1.1. City of Charleston Peninsula and the surrounding areas. ***City of Charleston Planning, Preservation and Sustainability Office***

GREEN LIFESTYLES

As we reach the ecological limits of the planet, increasing numbers of people are concluding that traditional environmental policies and activism do not work (Shellenberger and Nordhaus 2005). In this book, I show that discontent creates an opportunity for social innovation where change agents can develop, select, and enact alternative lifestyles. I document how everyday practices, like parenting and commuting, can potentially start a ripple effect in which greater, and ever wider, circles of change become enabled at all levels of individual and social life.

Green lifestyles are a collection of socially contextualized practices that people use to address challenges like pollution, climate change, and other environmental issues. Our lifestyles concern how we organize and carry out our daily lives. Within these lifestyles, we connect with others, differentiate from others, and make decisions. It is in our everyday lives that we make choices about things such as how to travel and what kinds of toys to provide our children. In this way, our lifestyles can have important impacts on larger systems, like the ecosystem (UNEP 2010). Sustainable lifestyles are an increasingly common response to the growing evidence that high-consumption living, especially in places like the United States (Global Footprint Network 2017), negatively impacts the environment (Cox and Cox 2016; Geisler and Currens 2017; Jamail 2019; Kalbar et al. 2018; Kolbert 2015; Matustik and Koci 2019; McKibben 2019; World Economic Forum 2019). Pope Francis (2015, part 206), for instance, focused his 2015 Encyclical Letter on global environmental problems, including climate change. The Letter is empowering and calls on everyday people to think about how "a change in lifestyle could bring healthy pressure to bear on those who wield political, economic, and social power." This book analyzes interview data, with parents and bike commuters, and draws on my own experiences as a parent and cyclist, to explore the everyday practices of people living green lifestyles and makes the case that daily life is, as Pope Francis and others point out, an important and under-recognized site for substantive change.

Green Parenting in the Lowcountry

Many of the forty-eight parents (GP) talked enthusiastically about how green resources in the Lowcountry are improving, including growing numbers of community gardens (two), nature-themed play groups and camps (five), and solar and alternative energy suppliers (twenty-four). A majority of the respondents (thirty) also brought up that, in the suburbs, there is an abundance of space for outdoor activities like food gardening and playing in neighborhood yards. A quarter (sixteen) pointed to the regional luxury of having abundant access to beaches, parks, and other outdoor spaces.

In spite of these amenities, green living in a red state can be challenging. Forty percent of parents made direct references to the conservative culture, like Rebecca (1 child) who left a Northeast corporate job to work on her Master Gardener's certificate in Charleston. She pointed to stigma she faces when people discover she works as a food gardening educator and explained: "I can do this without the stigmatism of 'okay, this is an alternative lifestyle' . . . I think we need to make this a little bit more mainstream, to where it's not extreme" (I-36, GP).

Avery (2 children) reflected on her upbringing in Charleston and told how:

> When I was little, I tried to start a recycling program, because it wasn't a thing. My mom told me that was ridiculous, and I shouldn't go through other people's trash. She was really upset. I was raised in a very conservative household, I had oddball ideas; then I went to [college at] Appalachian State. I discovered there were lots of people like me. I really, like, dove into it. (I-35, GP)

Later she talked about how she and her husband moved to Asheville, North Carolina, where, even though they were outdoorsy environmentalists, they were teased for being too preppy. In her words: "We didn't have dreadlocks and tattoos, smell like patchouli. We didn't go to yoga. . . . Then we moved here (back to Charleston) and all of a sudden everybody's like, 'you're a dirty hippie.' . . . It's such an odd transition because we've never been the dirty hippies of our group. We've been the least hippie [laughter]" (I-35, GP).

Difficulties with being "green" in a "red" state also emerged when parents voiced how they felt they could help others. Aware of their liminal status as green practitioners in the Lowcountry, 71% mentioned leading by example, and 50% emphasized that they avoid telling others what to do. Keenan (1 child), a kayak guide and naturalist, put it this way: "[I] lead by example, and doing the right thing, and people might notice it. . . . There's a fine line in the service industry . . . where you can offend certain people that don't [share your beliefs]" (I-47, GP).

His quote stands out because, even though Keenan works in a teaching field, he remains careful about what he says. Like him, Roselyn (1 child) adopted subtle approaches to influence others. In her words, she tries to take the tactic of "not being pushy, but just, this is what we're doing. This is what I'm trying to introduce to our life" (I-17, GP). Rosy (1 child) joked about how she feels that she sometimes says too much: "I talk about it a lot, probably more than I should. I try to plant seeds in people. I told all these people at work 'I'm gonna do cloth diapers' and they're like 'You're crazy.' Then

I'm like 'look it works.' I changed his diaper [while at work] one time just to prove [it]. . . . Maybe they will consider it later on" (I-19, GP).

While most parents avoided direct action, they did express a hopefulness that their efforts would eventually have a positive impact. In Avery's (2 children) words: "I don't know if my job is to fix the problem. [But] if I can be a good example, hopefully we'll all be thinking about . . . trying to live a different way" (I-35, GP).

These findings are especially interesting in that parents were not specifically questioned about Lowcountry culture; they were only asked to discuss the challenges that they faced.

While parents did discuss avoiding direct confrontation, their behavior should not be viewed as complacent; rather, they were adjusting their lifestyles as a method for supporting long-term systemic change. In her study of green lifestyles, Janet Lorenzen (2014a), likewise, found that her informants, rather than alienating listeners with overtly political talk, focused on being good examples to others. She concluded, as I do, that her informants were savvy about promoting environmental behaviors in their focus on being role models.

Lowcountry infrastructural and programmatic challenges that parents pointed to included: substandard walking, biking, and public transit infrastructure (thirty-five); not enough green- friendly homeowners' associations (HOAs) (seven); not enough recycling opportunities (five); and not enough daycare centers that accommodate cloth diapers. For instance, Lucy voiced her frustration with her HOA: "You're fined [for not watering your lawn]. . . . I feel so bad just using water for grass. It's grass. It's not a garden. It's grass. Here, the HOA, everybody has to have a green lawn." She went on to lament that "some of my neighbors, in the summer, it can be pouring rain, and then the next day they are watering their lawn because they don't pay attention. It's just that 'on' is automatic and it drives me insane. It's the same for the [whole] neighborhood because the neighborhood has irrigation everywhere and, of course, no one shuts that [off]."

She then compared her experiences with the Charleston HOA and her experiences growing up in France: "I'm from France and my mom always had a clothesline outside and, here in the summer, it's so hot. I'm sure you could put your clothes out there for ten minutes, but we can't. We can't do that. It has to be all done in the dryer" (I-7, GP).

Bike Commuting in the Lowcountry

Like parents, commuters also talked about the many amenities that the Charleston region increasingly offers. As I outline in chapters 2 and 4, all of the choice (CB) and many of the necessity commuters (NB) had at least one positive story to tell about the beauty of riding or the joy of connecting to nature in Charleston. In terms of regional improvements, respondents referred to a number of projects that are in the planning and proposal stages. For instance, there is the Battery2Beach initiative that is working to connect Folly Beach to Isle of Palms beach through proposed paths and lanes. The initiative is working on incremental path installation and getting regional ordinances passed to zone roadways for multi-use (Charleston Moves website). Several commuters voiced hope that the city's commitment to build a bike path over the Ashley River will eventually come to fruition. Eighteen choice commuters pointed to how much of a positive difference the building of Wonders Way path, over the Cooper River, made for bike commuters and pedestrians. Vince, for instance, recounted that the pivotal moment that got him bike commuting was the opening of the Wonders Way path: "The day the bridge opened; I was able to complete the loop to get downtown. Because, otherwise, I would have to try to illegally bike [Highway] 526 all the way around [to enter the peninsula from the opposite side] and I wouldn't do that" (I-6, CB).

In spite of these current and future cycling amenities, all of the bike commuters in both studies revealed concerns about current risks associated with the built environment.

All of the choice (forty-one) and almost half (nine out of twenty-one) of the necessity commuters made it clear that the local built environment poses numerous risks for cyclists. High speed and dangerous roads, difficult to navigate bridges, and the lack of bike lanes were most often brought up. Hudson, a choice cyclist, talked about how Charleston needs to build more bike lanes. In her words, this would make it so that drivers "have no choice but to respect you because it is there, you know? You can tell somebody to be nice. But if they don't want to be nice, they are not going to be nice. But if it is a law, like you see this space as bike space, and [you know that] you will get a fifty-dollar ticket [for driving on that space], you will not go on it" (I-4, CB).

Walter, a necessity cyclist, echoed Hudson's sentiments:

> Yeah, I want to add this right here. Here in Charleston, South Carolina, they got to give an area for bikes. You know, put that line, this is a bike spot, because when you riding down that Meeting Street, or whatever, they don't care. They'll run you over, or they'll hit you, they don't care. They need a space for bikers. And, also, bikers shouldn't be on the sidewalk, 'cause you zipping in and somebody come out the store, bam, you know, you hit them. . . . [Charleston has to] be better, there are a lot of bikers here, I see nothing but bikes all over Charleston. They got to make space for the bike. . . . They got to probably widen the street, bust the sidewalk down, put a little ten-feet or five-feet [path]

> for bikers. You know, on both sides of the street. . . . This way there won't be accidents. (I-8, NB)

Rodney pointed to another element of infrastructure that is lacking in Charleston. Many necessity cyclists combine bus riding with biking. However, the bus system (CARTA) is slow, unpredictable, and inefficient. Also, as Rodney pointed out: "I use CARTA. I don't know why they don't have bicycle racks on the back of the bus. You know, they got 'em on the front, but they could put 'em on the back and that, that would give two more people room for their bikes, you know" (I-18, NB).

Bike commuters, like green parents, also indicated that the cultural climate of Charleston is challenging. Thirty-five of the choice and eight of the necessity cyclists pointed to an anti-bike culture that presents dangers to bike commuters.

Choice cyclists linked bike-riding risks directly to the local driving culture even though most of them own a car. They think drivers see themselves as being the insiders and bike riders as outsiders on the road. This makes them fearful that, because drivers see them as outsiders, they can be justifiably vindictive toward them. They referred to a host of behaviors and comments of drivers as evidence. As one example, choice cyclists cited that drivers yell at them to "get off the road" or "get on the sidewalk" revealing, for them, a belief system among drivers in which cars belong and bikes intrude. They articulated that drivers see cyclists, the outsiders, as responsible for any risks that ensue from cars and bikes sharing the roads, and that the cyclists, therefore, should reduce those risks by getting "off the road." Michael, a choice cyclist, even reported intimidation from police officers, in his recounting:

> I was riding on King Street one evening. . . . I had a lane, and you know, if you're in a lane, you're allowed to be in a lane on a bike, you're going with traffic. And I had a cop pull up behind me . . . and do the loud horn, have you ever heard their horn? It's like they have that one horn, it's not like a car horn, it's like a siren horn or something . . . scared the hell out of me. He didn't pull me over, he just honked and drove around me. (I-32, CB)

In one case, Daniel, a choice cyclist, referred to symbolism inscribed in the landscape when he described how unwelcome he felt on the road after seeing some anti-bike graffiti. In his words: "Out on John's Island there's a 'Share the Road' sign with a big circle and slash through it. It just doesn't give you a good feeling. You know, you could be hunted, theoretically. If you've got time to do that, it's like wow, you went out of your way to hate me" (I-12, CB).

Crystal, a choice cyclist, employed a class-structure analogy to explain: "It's a general lack of consideration. . . . People here think that cars have the right of way over pedestrians, over bicyclists, so there is an attitude of car drivers, like *this is my street, roads are built for cars*, and bicycles and pedestrians, we are *second class*" (I-35, CB).

When asked how cars demonstrate a lack of consideration, she replied: "If I have a green light to go straight and I have a car that tries to turn in front of me, then that shows that they think it is their right of way instead of letting me go straight through the green. . . . [And] I do have people yell at me out the car window to 'get on the sidewalk.' I had people yell . . . [insults] at me and stuff like that just for being on a bike on the road" (I-35, CB).

Eight of twenty-one necessity cyclists also brought up Charleston's anti-bike culture. Jay put it this way: "People ain't gonna really pay attention to you. . . . They gonna blow you out the way" (I-6, NB). Anthony explained that he "almost got hit by a car a couple of times. They have no thought . . . no respect for bicycles. You see signs for 'share the road' but they don't always share" (I-2, NB). Likewise, Dean described intimidation he has received from drivers: "We don't get enough respect on the roads. You know . . . sometimes you can be riding down the road and somebody'll honk their horn just to try to scare you and stuff like that, you know. . . . We just don't have enough . . . rules and regulations, you know for bicycle riders. The cars should try to respect us more" (I-16, NB).

One necessity and thirty-nine choice cyclists reported that, in addition to outright harassment, there is simply a lack of attention to the bikers on the roads. In the following quote, Todd, a choice cyclist, discussed how poor driver behavior is a result of both ignorance and, sometimes, enmity toward bikers: "It's an everyday occurrence [in which someone] might be driving too close because they don't have an appreciation for the size of their vehicle or it might be somebody, you know, flipping me the finger, it might be somebody honking at me, it might be somebody swerving at me" (I-23, CB).

Carlton, a necessity cyclist, told the following story about an accident related to a driver's lack of awareness:

> Yeah, one evening, I was riding my bike down Rutledge Avenue and a car opened, someone just opened the driver side of the door, without even looking to see anything, and I just ran into it. I damaged more of his vehicle than he did my bike. All I did was broke off the brake handle. But his whole door was messed up. Glass broken out. But I was on the ground. I took the fall. And there could have been a car coming behind me and coulda ran me over. (13, NB)

Like parents who voiced that they want others to view their green behaviors as *normal*, one necessity and ten choice cyclists also explained that they feel stigmatized and want their activity to be seen as *normal*. Fran, a choice cyclist, told me: "I'm older. I'm overweight. I'm not a big spandex-wearing biker." She has been researching the purchase of a commuter bike called "Electra Tandy" and talked about how she liked that the company advertises the brand as "no strange clothes. No Lycra" (I-1, CB). Mark, Miles, and Jay echoed Fran's sentiments in the following quotes:

> I've never been . . . a gear-head type. I've always just kind of used whatever bike. . . . If I smoked cigarettes, I'd smoke on my bike. I don't like defining myself as a bicycle rider 'cause it's just how I get from one point to another. . . . It's people like me, not that I am special, but it's the opposite . . . regular people who don't want to be fanatical bicyclists, they just want to use their bikes in an everyday kind of way for whatever reason. (I-2, CB)

> I love briefcases on bikes. I love lights on bikes. I love fenders on bikes. Because riding is not about . . . clicking into your pedals. (I-40, CB)

> [People need to see that] . . . a bicycle done became a part of people's lives now. Maybe not everybody, but the biggest majority. I feel like on the bike, them became a part of a way a person lives because it's a big part of transportation, within they lives, you know. (I-6, NB)

CONCLUSION

Green parents (GP) and bike commuters (CB, NB) navigate a unique Lowcountry culture to make sense of and enact change in their everyday lives. While both parents and commuters pointed to the challenges of practicing sustainable behaviors in a conservative and/or anti-bike regional culture that lacks adequate green infrastructure and programs, they all persevered. For bike commuters, the infrastructure and Lowcountry anti-bike culture, discussed in this chapter, make up the majority of their challenges. In chapter 3, I outline additional challenges that are unique to green parenting. In the conclusion, I discuss how programs and policies could be made more successful if they work to unpack, understand, and address the unique challenges that all varieties of green practitioners, living in different social/political/geographic regions, experience.

NOTES

1. I am referring to a somewhat recent trend in which anti-environmentalists modify their trucks to spew plumes of black and smelly exhaust into the air. Often the smoke is aimed at pedestrians, bikers, and electric cars.

2. Charleston County has a median annual per capita income of $43,141 (United States is $35,383) and a median household income of $67,182 (United States is $64,994). Twelve percent of Charlestonians live in poverty (United States is 11%). Ninety-two percent hold a high school education (United States is 89%) and 45% hold a bachelor's degree or higher (United States is 33%). In terms of race and ethnicity, Charleston has a large Black population and a low Hispanic and Asian population compared to the national average. The White portion of the population is lower than the national average. Seventy percent are White (United States is 76%), 26% are Black (United States is 13%), 5% are Hispanic (United States is 18%), and close to 2% are Asian (United States is 6%) (U.S. Census Bureau 2022b). The Hispanic population, however, is estimated to have grown by as much as 55% between 2010 and 2020, and may be currently as high as 7% of the Charleston population. Even this figure is relative as South Carolina draws both seasonal and undocumented workers from Mexico, and Central and South America, for agricultural and construction labor (Halani 2018).

3. Early research on race and pro-environmental beliefs suggested that U.S. Black and other minority citizens expressed lower concern than Whites for the environment (Gifford and Nilsson 2014). However, over the last couple of decades, research points to how Black citizens express similar (Parker and McDonough 1999) and sometimes greater concern than Whites (Mohai and Bryant 1998; Uyeki and Holland 2000) and that the early research measured pro-environmental beliefs in terms that were less salient for minority communities (Arp and Kenny 1996). In addition, racial and ethnic minorities do express high concern for environmental justice issues. Countless articles and books have documented environmental justice activism in the United States over the last few decades. For two of the most recent works, see Josiah Rector's and Tracy Perkins' books on regional environmental justice activism (Perkins 2022; Rector 2022). Finally, beliefs and actions do not always connect and a growing body of research documents people and communities that engage in green behaviors for non-environmental reasons (for one example, see Wilhoit and Kisselburgh 2017), like many of the cyclists in the two commuter studies analyzed in this book (CB and NB). It is important to note that the necessity commuter study (NB) was made up of a majority of Black and male participants, and they were all engaging in the pro-environmental behavior of bike commuting, though only one out of twenty-one respondents expressed pro-environmental motivations for doing so.

4. One study found that the relationship between gender and strength of pro-environmental beliefs varies by race and ethnicity. In their analysis of survey data from a random sample of male and female White, Black, and Hispanic U.S. residents, Linda Kalof and colleagues (2002) found that gender differences only existed for Whites and that White men scored lower than any other group on strength of pro-environmental beliefs.

5. The study was aimed at families who currently had or recently had young children in the household.

6. According to the Pew Research Center, in 2014, 41% of mothers reported having two children by the end of their childbearing years. The two-child household was found to be even more common among well-educated mothers, as 46% of women

with a bachelor's degree reported having two children (2015) by the end of their childbearing years. While only 35% of the mothers in this green parenting study had two children, many of them were not at the end of their childbearing years. Eighty-one percent of the interviewees were in their thirties.

7. The choice commuter (CB) study had the youngest participants in it; this reflects national trends in which younger workers are more likely to bike to work (Burrows 2019). The older profile of the participants in the necessity commuter study (NB) might reflect the relatively large numbers of participants in that study that were unemployed, on disability, or retired.

Chapter 2

Pathways to Living Green

As I recounted earlier in this book, I bike with my son to do chores, get an ice cream, or visit the basketball court. We have not always done this. For one, I did not have a baby seat for my bike. Even after Gaspard became proficient at cycling, it took me a moment to start this new routine. What helped me make the change was a conference trip to Copenhagen, one of the most bike-friendly cities in the world. Because it was so easy and I already loved cycling, and because I was able to draw on my former experience of bike commuting in college, I was able to spend the week cycling around the city. I came home with a reinvigorated set of bike practices. While I still cannot commute across all of Charleston, like I can all of Copenhagen, the experience did inspire me to get myself and my child on our bikes, out of our block, and down a few streets to an ice cream shop. As social practice research emphasizes, my biking practice evolved out of a significant break in routine. In Copenhagen, I was surrounded by others who were bike commuting, it was easy to model them, and I felt normal doing it within a physical and cultural infrastructure that supported my activity. Years of studying bike commuting in Charleston had not gotten me on my bike. Copenhagen got me on my bike. It was as though I "remembered" I could do it. As Doug MacKenzie-Mohr and William Smith (1999, 95) write, the most important "influence on our attitudes and behavior is not the media, but rather our contact with other people."

Tracks to green living are made out of many overlapping experiences and beliefs. Sometimes there is a specific initiation. More often, like in my own personal narrative and as I illustrate in the stories below, practitioners recalled an evolving process.

PIECEMEAL PATHWAYS TO GREEN PARENTING

I allowed interviewees to define, for themselves, what it means to live green. Parents named a variety of activities, including low-intensity habits, like turning off the lights in the house, being moderate about air-conditioning and heat usage, recycling, carrying reusable bags, buying organic products, picking up litter, and shopping at secondhand stores. They also named high-intensity (involving significant time, money, or knowledge) habits like cloth diapering, making baby food from scratch, installing solar panels, installing a rain barrel, organic gardening/composting, updating insulation in a home, using public transit, and bike commuting, among many others.

The sample reflects a broad range of understandings of green living including one White single mother (Tammie, 1 child), who works in a factory, picks up litter with her child, recycles, and keeps plants in her apartment (I-6, GP); to one Black married mom (Germaine, 3 children) who earned her culinary degree as a single parent while supporting her children on public assistance and then started an organic urban farm and grocery store in a food desert (I-48, GP); and to one White married dad (Keenan, 1 child) who has a bachelor's degree in outdoor education, works as a naturalist and kayak guide, and makes most of his family's personal-care products, including chemical-free sunscreen (I-47, GP).

As I discuss in chapter 4, the forty-eight parents tended to already have green identities and then, with the birth of their children, intensified their commitment to creating a sustainable future by modeling green beliefs and behaviors. Forty-three of the parents talked about how becoming green was a piecemeal process. Their practices tended to begin with one simple act, like seeing another way of doing things while traveling abroad. The one behavior, or attitude, led to others over time. Among social practice theorists, this is referred to as positive spillover (Nilsson et al. 2017). As Rose (1 child) explained: "I've been headed that way my whole life" (I-19, GP).

Sharice (2 children) recalled in more detail:

> I don't think there was a turning point. It was just a way of life . . . I didn't realize [my parents] were green at the time. It wasn't trendy. But now, looking back, I just laugh because everything we're supposed to do, we did it. . . . It was part cost, like not having a ton of money. . . . My dad grew up on a farm, so he was familiar with that kind of thing. So, his mother always canned and froze vegetables, which to me is part of sustainability. My mom picked that up. . . . We recycled before they even had colored bins. My dad drove to a recycle park and got the money back. (I-18, GP)

Lucy, similarly, explained:

> It was no defining moment. It's just what I was raised to do and what makes sense. It's just thinking about the future instead of just thinking about the present. I don't feel like it's a huge change in life to just recycle and compost. It's

> not that hard. I don't know. I never feel like "oh, I live a green life." It's not something I think about at all. It's what we do. (I-7, GP)

Only five parents named one specific key moment that led directly to their green living. Abby (2 children), a vegan chef, for instance, talked about how, when she was 15 years old, a light bulb went off when one of her friends "gave me a video to watch, *Meet Your Meat*. . . . I never ate [meat] again" (I-10, GP). Similarly, Lindsey (1 child) reflected that "we grew up very typical. Actually, my parents were not very conscious. . . . In our adult lives . . . when we moved to Charlotte . . . Mike was trying to get on the Fire Department. While he was waiting, he started working at Whole Foods. That opened up the education component for us" (I-11, GP).

What mix of experiences pointed the forty-eight green parents to sustainable living? The vast majority came to green living through family and close friends (twenty-nine) and travel and exposure to other cultures (twenty-seven). Many also came to green living through healthy eating and habits (sixteen), outdoor experiences (fifteen), school and education (twelve), and media (ten). Other less dominant pathways included: new job experiences (three), moving to a new neighborhood (two), frugality (two), and concerns about animal rights (one).

Family and Close Friends

Sixty percent of parents pointed to experiences with family and friends as pathways to green living. Rose (1 child) recalled: "My parents were like, ok, we'll have a garden and we'll learn about the trees, and learn about the birds, and the outdoors. Then I was like, hey, I like the trees, and the outdoors. Maybe I should try to protect them and maybe I could consume less to help sustain the nice things that we do have" (I-36, GP).

For others, it was exposure to a companion, later in life, that helped them gradually lean into green living. One dad, Keenan (1 child), indicated that his wife was an important source of inspiration for his own engagement in sustainable living and that "she's studied environmental science and things like that too. So, she helped give it a step up" (I-47, GP). I interviewed him at a café in downtown Charleston. As we drank coffee out of ceramic mugs, he fed his daughter a fresh avocado, still in the skin, that he had brought with him. Like Keenan, Emma (1 child) talked about how she began living green when she "started dating [her] husband. . . . He started to get into health. Then we just got into it together and started restructuring healthier ways of life" (I-34, GP).

Travel and Exposure to Other Cultures

Fifty-six percent pointed to the importance of travel and the exposure to other cultures. Jane (1 child) talked about how, as a young adult living in California, "I was always keepin' up with the Joneses. Right? I had to buy those jeans that were $300. [Later], I lived in Peru for three years. My values changed. There's nothing you can buy. You can only have so many hemp bracelets. You know what I mean? It was amazing the change in me and the way my body felt" (I-23, GP). Jane then brought up that she had just received her tax refund and was suddenly thinking of all the things she wanted to purchase; she stopped and told herself: "No, remember Peru. You need nothing and you feel better with nothing."

Another mom, Cab (3 children), noted that her green living evolved out of a number of experiences, including that she traveled

> to Europe and I understood from an international perspective how crowded the world is and the different choices made about transportation. Also recognizing [that] people live in small places. They don't have McMansions or a lot of clothes and that's great. They buy their groceries at a corner market and come home. That was really where my thinking changed. The first time I went to Paris, Rome, Germany, and especially Switzerland, I had a feeling of [Americans are not] very simplistic and that's probably [a bad thing] and my thinking changed about my footprint and the impact we're having on the world. (I-2, GP)

Likewise, Luna (2 children) talked about how:

> I keep conscious, I think, because I grew up in Peru. I would say things are more local or organic. If I eat something from a can in my life, it was tuna is the only thing, because . . . to get something in a can is expensive. . . . We always made everything natural. So, when I moved here, I would say, "oh, these don't taste as well." . . . I started seeing all the things they put in food. (I-8, GP)

Some parents even pointed to how they seek out travel experiences as ways to educate their children about green living. For instance, Elizabeth (2 children) discussed how her family had taken their children to Italy. "It was fun showing them what public transit is like. We didn't drive. We got on buses and subways. They loved it. I think that was important. 'This is how it is around the world, and things are different depending on where you live.' Just seeing some of the schoolkids on the subway, and things like that. The girls were fascinated" (I-38, GP).

Healthy Eating

Thirty-three percent of parents pointed to how their efforts to eat healthy and to live healthier lives led to other green activities and beliefs. Germaine (3 children), who now runs an urban farm and grocery store, discussed how food and health were a big part of her embrace of sustainable living. In her words:

> Well, affordability is my word. The initial catalyst was the fact that we got to a point as a family, when I first moved here, where I couldn't afford to feed my kids while they were young. I was a full-time student. . . . Immediately we had a real need for food. I had gone to pantries, waited a long time to even get a bag of stuff that I . . . don't even want to serve to them. That got old real fast. I just made it a goal that once we moved out of public housing, we'd have a house, and the house would have a garden. We did that. A lot of it is about being healthy and providing healthy food choices for your children. . . . A lot of the green habits, composting, all that other stuff, came out of that. (I-48, GP)

Another mom, Eliot (1 child), talked about how eating organic, unprocessed, and local foods fueled her evolution into green living. Once her family started eating a healthier diet, "everything else fell [into place]. We are way more responsible now about recycling, about using reusable shopping bags, about making more of the effort to go to markets and [having a meal plan]" (I-43, GP).

Outdoor Experiences

Thirty-one percent of parents pointed to outdoor experiences as important pathways to green living. Avery (2 children), when discussing how an evolving love of nature led to her current green lifestyle, said this: "When I grew up here [in Charleston], girls didn't surf. We hardly ever went to the beach. It just meant we were laying out in the sand. . . . [When] I got up to Asheville [for college], I was like 'Oh, people go outside all the time.' I became a climber, and I climbed almost every day and worked for an outdoor program. . . . That changed my life, amazing" (I-35, GP).

Another mother, Kareena (2 children), told the story:

> I am from West Virginia, and my father, both my grandfathers, were in the coal industry. I feel very strongly about the mountains and landscape. I became involved with the anti-mountaintop-removal coal mining movement a couple of years ago and [am concerned about] coal, fossil fuels, and carbon emissions on every level. Really, I think that's where I would like to decrease my footprint. . . . Because I have been personally affected by it in several ways. . . . My father was killed in the coal mines so it's very personal. I also feel passionately

> about the outdoors and how beautiful the mountains are. [It hurts] to see them destroyed [by] coal mining and fossil fuel. . . . It's visceral for me. (I-29, GP)

The two dynamics (family/friend influences and outdoor experiences) worked in unison in many of the parents' lives. Participants recalled how early family experiences, like venturing out together to garden or to go camping, made green living seem *natural*. It was just the way to be. April (1 child) explained that it felt normal to live green because when she grew up

> there was always a garden. I remember spending summers a few weeks with my grandparents, my mom's parents, and they always lived off the land. They grew everything—watermelon. I remember a big plum tree and sittin' under there eatin' all the plums. [We had] pecans and chickens. [My grandmother] would decide like, "oh, you want chicken today? Go kill your chickens!" I remember canning. My mom and grandmother sittin' out under that tree and canning and getting everything prepped to start canning all the winter food. (I-41, GP)

Another mom, Ray (1 child), who is now a commercial chemist and referred to herself as a volcano-obsessed scientist, recollected that, even though she did not like family camping trips when she was young, she sees now that she got her love of nature and eagerness for green living from her father.

> One of my favorite realizations was [when I] had a conference . . . in Portland last year. . . . I got very excited because Mount St. Helens was like an hour and a half drive. . . . I stayed an extra day [and] took a trip to Mount St. Helens. [I] stopped at some waterfalls and other things in the area to go hiking. . . . It wasn't even till the end of the day that it occurred to me that that's exactly the kind of thing my dad used to do when he would go on business trips. He would . . . would fly to Wyoming for something and go to Jackson Hole. . . . He'd go to Tennessee for something and stop at Smoky Mountains National Park. Always in parks, always in nature. (I-37, GP)

Rebecca (1 child) recalled that she had also been brought up to live green and love the outdoors, but that when she worked in the corporate world she got away from this way of living. She had a turning point while living in Washington, DC. In her words:

> Living in D.C., it was all the rat race and everybody's so important, and everybody's gotta have the designer stuff. One day I was just "so what is this? Can't do this anymore." That's when I [got] disgusted with how everybody was throwing money away and not really caring about anything else. . . . I had gotten so far removed from [how I grew up] and I came back to it and realized that that felt more like me. (I-36, GP)

She subsequently quit her corporate career to earn her Master Gardener certificate and now works with a farm-to-school community education program in Charleston. (I-36, GP)

Many parents expressed self-awareness that their childhood exposure to nature positively impacted their love of the outdoors and that they mindfully prioritized getting their children outside for that very reason. One parent, Eliot (1 child), noted: "[Getting outside] is very important to me. Not in the summertime 'cause it's so damn hot, but we ride bikes a lot. We go down to the beach a lot just to walk and pick up shells. We collect what we call our treasures. We go to James Island County Park a lot. Getting her involved in that aspect, I think, makes children, as a whole, and adults too, more appreciative of nature" (I-43, GP).

School and Education

Fifty-seven percent of parents talked about how school and education were important conduits to green living. Often, education, like experiences in college, stoked a love of nature that had already been fostered by outdoor family experiences in early years. Stella (1 child), who grew up in St. Louis, recalled: "My dad, when I was about three or four, . . . bought property an hour and a half west of where we were living, which to this day, we still call 'the land.' It's just land that we would go camp on or just go out for the day and spend time playing in the river. Eventually, he built a lake out there for swimming and fishing. It's just this weekend spot" (I-40, GP).

She then talked about how, though her love of nature was a gradual process rooted in early family experiences, later in college she learned more about environmental problems, such as pollution and environmental racism. Though her parents had instilled a love of nature, she explained:

> I grew up in the Midwest in a Catholic family, very, very sheltered. We just didn't learn anything else but Catholicism and how to be a good person. Even in high school, we just never learned about anything controversial. It was just, "read those books and learn this math, but we're not gonna discuss world issues at all." When I got to college and met all these more worldly people, I was just, my mind was blown. (I-40, GP)

Informal education through media also served as a pathway, for many parents, to green living. Though, as I will discuss in chapter 3, parents' relationship with media was complicated.

Media

Forty-seven percent of interviewees pointed to the importance of exposure to green media. Lyndsey (1 child) discussed the impact of documentaries and books on her growing environmental awareness. In her words:

> I know the last, let's say five or ten years, there's been tons of documentaries and books, and not just about some big scandal. It's more like everyday environmental problems, think about where your food's coming from. Think about where the trash is going. There's been more awareness . . . things like Netflix, online media, it's just so much more accessible. . . . It's easier to be aware. (I-11, GP)

Luna (2 children), likewise, talked about how she had grown up with parents who grew food and made a lot of their own products but that she is even more environmentally aware: "Now with all the news coming out about being conscious about the environment and the animals being in extinction . . . the chemicals that we use [in our] hair, cleaning products . . . everything that really changes the atmosphere. . . . You hear all this stuff on the internet" (I-9, GP).

Other Experiences

Parents brought up a few other experiences that led to green living. Three parents (0.06%) named new job experiences, two (0.04%) pointed to moving to a new neighborhood, two (0.04%) explained that frugality inspired many of their early green decisions, and one (0.02%) talked about how she became involved in animal rights and that this experience led her to other green behaviors. April (1 child) talked about how frugality was an important motivator for green living along with self-directed exposure to media about simplicity and, also, exposure to important role models and cultural practices.

> Going back to grad school, you're like, "Oh shoot, we don't have money!" What are some ways we can save on products? I can make certain things, and not havin' to buy a whole bunch of different cleaning products. I think, even at one point, I made laundry detergent, Borax, and some other stuff. [Then] when we were in the [old] house, we weren't in the city zone, so the recycling wasn't, we had to actually make an effort to recycle, to take it to the recycling places, and things like that. We did that because we were paying for a trash can . . . so it reduced the amount of cost. . . . I'd go to the library and get different books. . . . It's always like, financial literacy. . . . I cloth diapered the first eight months. I think it was a combination of things. My mom cloth diapered us, had a diaper service. I think it was more for economic reasons and because I think part of the mindset of growin' up too, like, it was, "you cloth diapered, right?" [laughs]. I

> cloth diapered the first eight months. I had a diaper service [and] there was this great cloth diapering store in Omaha. 'Cause, people out there, it's normal to have four and five kids. Cloth diapering was big out there. They gave classes on it. It was a lot less intimidating on what to do and buy, and all that. . . . [Now] once I got back to working more hours, . . . we try to use diapers that are, you know, break down better in the soil. (I-41, GP)

PIECEMEAL PATHWAYS TO CHOICE COMMUTING

All of the forty-one choice commuters, like most parents, talked about how utilitarian biking evolved piecemeal from a mix of motivations and experiences. For instance, Ginny, when recollecting how she and her husband got started commuting, referred to exercise, health, liking the outdoors, self-sufficiency, environmental issues, and urban planning:

> Both of us really like to be active and both of us find it really hard to be active if it's not part of our daily life. . . . I'm the kind of person who would rather chop wood than lift weights at the gym. . . . I just want to use my body every day. So that's a big reason for biking and I love being outdoors. And then the other part is that just environmentally you don't want to use a car. Well, I hate driving, first of all. I hate traffic. . . . But I also just feel like, I mean I don't want to be corny, but if I have the power to bike why would I want to [drive]? So, if at all possible, we . . . ever since I was in college, I started biking a lot. . . . I guess when I had kids I was bound and determined not to let that stop. . . . I guess [when] a city is like bike friendly, I think it changes the whole culture of the city. I think people are healthier, I mean people become fit, there is less pollution, there is less congestion, there is less road rage. I just think it is better for the whole community. (I-7, CB)

The forty-one choice cyclists identified a number of interlocking paths to commuting. Over half (thirty-one) explained that their commuting evolved out of recreational cycling, that they were already doing, and the desire to get more exercise. Many (thirteen) also pointed to how infrastructure improvements or moving to a more bike-friendly area enabled them to start commuting. The same number (thirteen) explained that they started bike commuting to put their environmental beliefs into action. A few (seven) talked about how they realized it was a pragmatic option after seeing other people in the community doing it. Similarly, a few (six) also talked about seeing family and friends as role models for commuting. Some (six) recollected that, though they can now afford a car, they got started out of necessity.

Exercise and Recreation

The majority, 76%, reflected on how recreating and/or exercise were key pathways to bike commuting.

In the words of Paris: "I guess I've sort of always been into riding and one day I kind of said, hey, why don't I ride my bike to work, just sort of as a, let me try it see how it goes. And then I realized how much I enjoyed it and it's like, okay, I think I'm gonna start doing this on a regular basis" (I-11, CB).

Often, interviewees pointed to joyous experiences that went all the way back to childhood. The following quote from Michael is typical: "You know I love riding my bike. . . . [As a kid], I was in a little neighborhood bike gang where we built ramps and rode around . . . so it was natural for me" (I-32, CB).

Crystal recalled biking as a young girl and told a story about how joyful childhood feelings were conjured up one day while cycling in a Charleston rainstorm:

> The streets flooded and I was riding back from yoga, and I have on my rain pants, and my rain jacket, and I'm driving down this street, and the water on Ashley Avenue is up to my pedals. So, when I approach a steep puddle, I put up my legs and try to coast through it without getting the water in my shoes . . . and it's so fun. It makes me feel like a little kid. . . . All these cars are going 2 miles an hour through the rivers, and they are so cautious and I'm going faster than them peddling through the river. And it's kind of like, ha ha ha, I'm out riding my bike in this, it's fun! (I-35, CB)

For many, recreation and commuting became so intertwined over the years that they could not be separated. For instance, Miles explained:

> I never did not ride my bike to where I was going. . . . I either walked or rode my bike to junior high school. And then I either took the school bus or rode my bike to high school. . . . In law school, I rode my bike, and I didn't have a car. After law school I lived in Europe for three years, and I didn't have a car because I rode my bike everywhere. It wasn't until I was thirty and moved to Charleston that I bought a car. . . . It wasn't really a choice not to buy a car. It just wasn't a priority of mine. (I-40, CB).

The desire for more exercise served as an important motivator for cyclists to transition from recreating to commuting. Bee explained that she had been a longtime triathlete and then had her first child. She realized, "I wouldn't have time to train at all unless I biked to and from work . . . and that was why I started doing it" (I-26, CB).

A part of what explains the large proportion of commuters that came from lifelong bike recreating is that their experiences made utility cycling feel "normal." Gunner, who had been riding a bike since he was 14 years old, reflected on how his childhood and adult history with recreational biking is key to understanding his ease with commuting as an adult. In his words:

> That is part of the formula. . . . I think there are key things that all have to line up. . . . You have to feel confident about riding a bike and I think a lot of people don't. A lot of people are very afraid about safety . . . and I think that is absolutely reasonable. I think that [there is] the concern of . . . if I get halfway to work and end up [with a] flat, am I going to be late? . . . If it rains in the afternoon, what am I going to do? (I-29, CB)

While most bikers came to commuting from lifelong recreation, some had to overcome the challenge of learning how to do it. Fran, for instance, described how

> [it] was a whole paradigm shift . . . having to decide, ok, I'm not gonna just ride around the neighborhood anymore so I need a helmet. Gee I better get a reflective vest. Walking into a bike store, which is a totally alien thing to me, and making sure I had lights. It was a total shift in thought making and [creating] time to do it. It's gonna take a little longer than driving. (I-1, CB)

She went on to joke that the bike shop does not seem weird anymore and "they're very helpful even though I go in and say, 'I'm riding a Wal-Mart bike'" (I-1, CB).

Infrastructure

For many (thirteen, 32%), the transition to bike commuting began with a residential and/or career move. Margaret recollected that she moved downtown and started driving to work and then it occurred to her, "Why am I driving to work when I only live two miles away?. . . . Now, I am stuck on it" (I-13, CB). Similarly, Colby said that he started bike commuting after he got a new job that was only three miles from his house: "So I'm like, I can do this. And I weighed two hundred and twenty pounds, so I could do it just barely." He explained that he had been trying to lose weight but had plateaued and "needed something to get me over that plateau and down a little bit." So, he recounted that he went into his garage and pulled down a thirty-year-old bike, dusted it off, took it to the shop across the street from where he lived, and started commuting (I-31, CB).

Several interviewees pointed specifically to the opening of the Ravenel Bridge *Wonders Way bike-pedestrian path* as an infrastructural change that allowed them to start bike commuting in Charleston. For instance, Jenny, an owner of two sports-related businesses and an avid athlete explains:

> I guess what really got me started commuting was the opening of the Ravenel Bridge [*Wonders Way path*], because I live downtown [but] my stores [are over the bridge] in Mt. Pleasant. And there is no possible way I could have ridden my bike over the old Cooper River Bridge. . . . Once I had a way to get to Mt. Pleasant on my bike, then there was no question, I knew I was going. (I-15, CB)

Several noted that they got started commuting when they intentionally moved or changed jobs to have access to bikeable roads. Margaret, for instance, moved with her boyfriend, from Sullivan's Island to downtown Charleston, to live closer to work so that she could bike commute. In her words: "When we were looking for a place, I knew that I really wanted to be able to ride a bike so we had a few miles radius was our zone of where we wanted to live, and we were very happy to find this place so that I can ride" (I-13, CB).

Environmental Motivations

Thirty-two percent explained that they started commuting as a way to put their environmental beliefs into action. Gunner remembered that he started bike commuting for both environmental and fiscal reasons the year that *Exxon Valdez* crashed off the coast of Alaska. In his words, "I was a sailor then and I was stationed on a ship on the main base down here and . . . I decided I wasn't going to be part of paying Exxon's bill" (I-29, CB). Vince also referenced oil when he talked about his environmentally informed pathway to bike commuting. He recalled:

> I've always been concerned about the environment and conservation and being a good steward. . . . Then I was in Iraq being a doctor with the Marines . . . I really took an oath to be a conservation warrior when I got back because of the root causes of the war. It's pretty simple, if you are biking to work, you are burning calories, you are not burning oil. (I-6, CB)

Caleb, who gave up his car as part of his commitment to bike commuting, told that he got started for "a combination of environmental ethics and personal financial reasons. I was preaching a lot of environmental advocacy to my friends. . . . I felt like I was a hypocrite" (I-19, CB).

It is important to note that these "environmentally aware" pathways to bike commuting may be over-represented because I prompted the participants. One of my original research questions had to do with the degree to which environmental awareness informed a person's decision to start bike commuting. So, if respondents did not mention environmental reasons as a pathway to commuting, I would prompt them with questions like "were there any big picture reasons for your decision to commute?" Many responded with a conditional "yes." For instance, they would say "yes" but that they really started for the exercise first. In the section of the interviews on current motivations for bike commuting, covered in chapter 4, I did not prompt choice cyclists. If reasons for currently commuting did not come up on their own, I asked "why do you continue to commute and what do you like about it?"

Seeing Community Members and Close Role Models Commuting

Seeing others successfully ride their bikes for utilitarian reasons, especially family/friend role models, was an important avenue to commuting. Fifteen percent of respondents indicated that seeing other people commuting inspired them to take it up. Samuel, a longtime recreational cyclist, commented that he started bike commuting when he lived in Washington, DC. He was riding the subway to work but "saw a lot of coworkers bike commuting, so that's when I started really commuting, you know, where it meant carrying clothes in panniers" (I-22, CB). Mark, similarly, was influenced by the biking sub-culture at his workplace. As he explained:

> On sailing ships . . . when we would get into a port, if we had a bicycle on board, I would use the bicycle. . . . A lot of times shipyards had bicycles. It's very much a part of the manufacturing and ship repair industry because you have these sprawling complexes and it doesn't always make sense to get into your pickup truck. There is not always parking. . . . So, bicycles are a staple of shipyards. (I-2, CB)

Seventeen percent talked about the importance of close role models. Kelly recalled that she became a bike commuter, as a child, in the 1970s during the first gas shortage. As she recalled: "My dad decided he didn't want to spend money to put gas in our gas-guzzling station wagon and that he was going to take me to school and himself to work on a tandem bike. . . . I have very fond memories of going to school with dad on the tandem bike" (I-16, CB).

Arnie, who at the time of the interview had 50,000 miles of biking under his belt, recounted that in high school he had an economics teacher who started a bicycle club and that he had been riding hard ever since. The teacher bike commuted to school and successfully encouraged many students to do the same. He recollected fondly that he would be cycling to school and, sometimes, he would sneak up on his teacher and latch a finger onto his bike rack and get a free ride (I-5, CB).

Two respondents traced their bike commuting to the role modeling of friends. Caleb, an ecologist, had wanted to bike commute but was not sure how to transport his heavy field equipment. He became even more motivated after a friend sold his car as part of his own commitment to bike commuting. Then, when he verbalized interest in commuting, he had another friend, who worked at a bicycle store, explain how Caleb could haul his equipment by hitching a trailer to his bike. He also offered to help Caleb with repairs and equipment (I-19, CB). When driving around James Island (a suburb of Charleston), I sometimes see Caleb on his bike, with his trailer, headed to one of the local beaches to study wildlife.

Necessity

Finally, 15% of respondents indicated that, when they started bike commuting, they had no other choice because, though they can afford a car now, they could not at that time. Luca, for instance, noted that he did not have money for a car in college, so he biked instead. Many years later, when living in Houston, he owned a car but decided to ride a bike. He explained that driving a car never seemed like a necessity to him as he "grew up in Africa, where many people longed to have a bicycle and very few people had cars" (I-39. CB).

While some of the choice cyclists started bike commuting out of short-term limitations with financial resources, all of the necessity cyclists came to commuting from circumstances where they could not afford a car. In addition to not being able to afford a car, some had lost their driver's license. While none of the necessity cyclists could afford to drive, many of them, instead of talking about economic restrictions, pointed to other reasons for starting commuting.

PIECEMEAL PATHWAYS TO NECESSITY CYCLING

The interviews with twenty-one necessity cyclists were very short compared to interviews with green parents and choice commuters who I was able to recruit through traditional means with flyers, chain-referring, and (in the case of green parents) social media posts. Necessity cyclists did not respond to this form of recruiting. I adjusted by visiting places that necessity commuters frequent (public laundromats, libraries, soup kitchens, and popular street-corners, etc.). Necessity cyclists were, therefore, often in transition and not in a position to speak at length. Nevertheless, they told similar stories about the piecemeal pathways to their commuting. For instance, when asked how he became a bike commuter, Martin brought up both necessity and lifelong recreational biking when he answered: “I ain’t have no license . . . I been riding bikes since I’ve been small” (I-9, NB). Jacob referred to wanting to get to know the layout of the city better, and the time and cost savings. In his words:

> I always rode a bike but . . . since I’ve been here in Charleston . . . the main reason I got this bicycle (from the Salvation Army for $10.00) was to learn the city. Because a bus is taking you one route. But since I had this bicycle, I done learned about three routes to come down here from North Charleston . . . to get to the VA [Veterans Administration]. I done learned about four ways to go. Where at one time [on the bus] it would take me ’bout thirty minutes or more to get to the VA, it’s eight minutes tops . . . on my bike. So really . . . I wouldn’t take nothing for this bicycle. . . . You don’t have to buy gas for [it] . . . it’s manpower. (I-7, NB)

The twenty-one necessity cyclists pointed to several key pathways to commuting. Not surprisingly, nearly half (ten) indicated that they started commuting because they could not afford a car and/or had lost their license. Many (six) pointed to how they had simply always bike commuted recreationally and wanted the exercise. In addition, six talked about how convenience served as a strong motivator to get started commuting. One talked about how he began commuting for environmental reasons.

Necessity

Ten (48%) of the necessity cyclists shared that they got started commuting either because they couldn’t afford a car and/or had lost their license. As Arthur recounted: “Back in the ’80s, I had my own car, but, ya know, [I lost it] from a DUI charge [and] I really got in some trouble with child support and all this kinda stuff. I said, “The best thing you . . . just give me a bicycle ever since then” (I-20, NB). Tony recalled that he first started commuting because he lived in Baltimore and “so it didn’t make no sense to drive.” Now that he is living in Charleston, he explained that he could not afford a car, and even if he could afford a car, he could not afford insurance (I-5, NB).

Convenience

Twenty-nine percent explained that biking is more convenient than driving or taking a bus. Clark explained that convenience was one of the reasons he started commuting because: “I like the fact that I get to pass a lot of cars and I don’t have to worry about parking or gas” (I-12, NB).

Exercise and Recreation

Exercise and recreation (29%) were important motivators for commuting. When asked how he originally came to bike commuting, Alejandro, who had been cycling on and off since childhood, explained: "I come out of a coma. I couldn't walk. So, I built up strength [on the bike], you know" (I-1, NB). Often, necessity cyclists articulated the exercise/recreation origins of their bike commuting as having *always ridden*. Dean explained, "Ever since I've been small, I've been riding bikes, you know. It's a good way to get around, you know, without using a car and stuff like that" (I-15, NB).

Environmental Motivations

One participant, Howard, reported that he started commuting for environmental reasons. He explained it is "mainly because, one reason, is a political thing and it's educational because of the pollution. They have emissions control. You ever heard of emissions control? So that cuts down on pollution. . . . Bicycles are a cheap form of transportation [and] cut down on pollution" (I-3, NB).

CONCLUSION

A growing interdisciplinary social science literature asks: "Why and how do people come to live pro-environmental lifestyles?" (Colby 2020; Geels 2011; Genus and Coles 2008; Grin et al. 2011; Markard et al. 2012; Scurry and Bowden 2021; Siegel et al. 2018; Tamas and Bohm 2020). In an important review of decades of research on this question, Siegel and colleagues (2018, 189) document that the answers to this question remain debated and complex. They suggest that we think of "a *living forest* . . . " where a web of "soil, fungi, plants, insects, animals, water and air" cultivate green living. In line with this vision, parents and utility cyclists shared the common story that paths to green living were largely ongoing, overlapping, multi-faceted, and diverse. Practitioners came to green living through efforts to eat and live healthier; the desire to enact a sustainable lifestyle, economic necessity, and/or convenience; outdoor, educational, media, infrastructural, and travel experiences; and much more. Sometimes people come to green living with intention and other times it was an outcome of other desires, like wanting to exercise, or need, like financial concerns.

Echoing the *living forest* imagery, social practice theory (fully cited in the introduction), proposes that pro-environmental habits are dynamic and emerge out of everyday life in which practices consist of patterned relations between individuals, social groups, and the physical world and are made up of meanings, resources, and skills. For practitioners in this study, their myriad habits emerged out of meanings (that is, the joyous memory of biking as a child); resources (that is, land on which to grow food); and skills (that is, knowing how to go camping in the outdoors). An important part of the development of social practices is social modeling; participants in these studies pointed to how social modeling, especially that of close family and friends, but also community members, enabled their initiations into green living (for instance, childhood memories of bike commuting or canning vegetables with a parent) (Greenspan et al. 2022; Hendlin 2016; Wilhite 2015; Lorenzen 2014b).

The *living forest* also draws on the concept of embodiment which proposes that it is through our physical doings (behaviors) that we act on the world and, at the same time, create our own larger realities and possibilities (Averett 2021; Bashford 1998; Bennett et al. 2013; Benzecry 2009; Berger 2008; Brady and Phemister 2012; Carolan 2009; Cresswell 2006; Diprose 1994; Durt et al. 2017; Eicher-Carr 2001; Featherstone et al. 1991; Foucault 1979; Fox et al. 2014; Ingold 2004, 2011a, 2011b; Merleau-Ponty 1992; Messerschmidt 2004; Oliver 2020; Shilling 2007, 2004; Tarasoff 2018; Wacquant 2005, 2006, 2015). In the words of Michael Carolan (2009, 2) "we do not simply exist within a material reality. Rather, bodies help constitute that reality . . . through our doings within it."

An important subsection of this literature explores embodiment and nature (GhaffarianHoseini 2012; Humberstone 2010, 2011, 2013, 2015; Kidder 2006b; Nightingale 2012; Payne 2006; Twine 2001; Weiss and Haber 1999). Research on nature embodiment is diverse. Some studies have focused on instances where sustainable practices decay and/or fail to materialize, or where people experience challenges with connecting to nature. Andrea Nightingale (2012), for instance, drew on the embodiment perspective to explore the issue of what makes people work together to protect shared natural resources in some instances and not in others. Her study focused on Scottish fishermen and looked at how their understandings of nature emerged out of their bodily interactions with the communities, spaces, and places of fishing. She found that the fishermen engaged in boundary making as they went through multiple work-related transformations; for instance, they moved " . . . from their boats, to the pier, to the meeting room . . . " (135). In some cases, fishermen's attachment to the sea and community helped to provide motivation to engage in pro-environmental fishing practices. In other cases, the boundaries that developed, especially between places where they interacted with regulators and their own embodied experiences of fishing at sea, presented challenges to working together to protect the ecological environment. Other studies have focused on how positive ecological habits can evolve out of the sheer bodily pleasure of engaging in sustainable activities (say, the visceral joy of bike riding) (Kidder 2006b). Green parenting and bike commuting are embodied in that they are both about the goal, raising a healthy child or getting somewhere on a bike, and about the doings themselves (like the enjoyment of collecting shells on the beach with your children). The ultimate objective is not just a healthy child or a destination, it is to experience the felt sensation of parenting and/or moving on a bike. It is also embedded; one must navigate social/political/geographic contexts to physically care for a child or cross a landscape (Carolan 2011; Jones 2005; Ingold 2000, 2011b).

The themes of education, media, and travel also reverberated strongly through the narratives. Education and educational media have been well studied as precursors to pro-environmental behavior and awareness (see the following journals: *Australian Journal of Environmental Education*; *Environmental Education*; *Environmental Education Research*; *Environmental Education and Sustainable Development*; *International Research in Geographical & Environmental Education*; and *South African Journal of Environmental Education*). The positive environmental education impacts of travel and of mobility have been less studied (Brody and Ryu 2006; Flint 2001). Results from this study show that travel[1] is an under-researched, but important, pathway to green living. A few research projects have been conducted on the value of environmental education within tourism (Zheng et al. 2020; Ballantine and Packer 2011) and education-related travel (Tarrant et al. 2017; Stoner et al. 2014), but more needs to be done to study the value of travel, and also mobility (moving from one home to another, for example), for inspiring pro-environmental behaviors and beliefs.

In the conclusion, I will discuss the policy and programmatic implications of these findings, especially that the origins to green living are piecemeal and overlapping; that they emerge out of everyday socially contextualized practices in a constant feedback loop; that environmental habits are embodied and, often, viscerally pleasurable, and that place is important, as we *perform* our *green practices* in *physical* and *social spaces*; and that a host of experiences and interests can feed into pro-environmental behavior, like healthy eating, financial necessity, education, starting a new job, and so on.

NOTES

1. The finding that travel is associated with increased pro-environmental concern and behavior presents a set of contradictions that would have to be addressed. A large proportion of tourism and educational travel occurs by air transportation, which has been recognized as a major contributor to climate change. UNEP put out its most recent Green the Blue 21 Report (2021b), which documented that, in 2020, air travel

accounted for 32% of CO_2 greenhouse gas emissions. In addition, 25% fewer greenhouse gas emissions were released in 2020 compared to 2019 due to the impact of COVID-19 on travel restrictions, among other things.

Chapter 3

Green Parenting Challenges

In 2011, when I became a new mom in a dual-income family, I found myself searching for any technology, product, or advice that would help our household continue to live an environmentally sustainable lifestyle. However, I also wanted time with family and to pursue my academic career. Like the respondents in the parenting study, I found myself struggling with the financial costs of green living (that is, purchasing eco-friendly diapers and other products), the time to enact green living (for example, cloth diapering), and the difficulty of navigating green media (like the debates over cloth versus eco-friendly diapering).

Interviewees confirmed that the arrival of a child impacted their green lifestyles and they pointed to the following overlapping challenges: conservative Lowcountry culture and substandard infrastructure (previously analyzed in chapter 1); money (twenty-eight); limited time to get things done (twenty-eight) and investment time to learn a new routine (fifteen); and the insufficiency (thirteen) and/or overwhelming nature of green media (nine). In addition, parents brought up another conflict with media—the struggle and desire to limit family screen time and to re-emphasize time together and experiences with nature (fifteen).

MONEY AS A GREEN PARENTING CHALLENGE

Fifty-eight percent of parents listed financial constraints to green living. When they talked about money-related challenges, they referred to sustainable practices as a whole and not just the costs of raising children. Abby (1 child) explained: "If we had more money, then I can get the solar panels. If we had more money, then I could buy all the green products—recycled paper towels and things like that. We could buy organic and local fruits and vegetables" (I-10, GP).

Isabelle (1 child) also pointed to financial constraints. Her family rents in a sprawling suburban neighborhood, but their goal is to move to the more expensive and pedestrian-friendly Charleston peninsula. She explained, “West Ashley is super dangerous for biking and walking. If I could live in a walkable community, I would” (I-21, GP).

As evidenced in some of the quotes below, parents often brought up money and time together as comingled challenges.

TIME AS A GREEN PARENTING CHALLENGE

Fifty-eight percent of parents reported that time was scarce. Stella (1 child), is an extraordinarily committed green practitioner. She completed several organic farming internships in college and owns a business where she helps other families set up their own food gardens. Her husband also works in organic agriculture as a sharecropper on Charleston County’s Johns Island. They cultivate their own food garden at their home. She shared that they are so busy that, though they work in the business of growing local and organic produce, she usually heats up frozen meals from a big-box store for their work-night dinners (I-40, GP).

Parents often expressed that time and money are parts of the same problem, like Jane (1 child), who explained: “I’m a schoolteacher. I’m a one-income schoolteacher” (I-23, GP). Jane discussed how she would like to have solar panels but couldn’t afford it and that, because she is a working single mother living in an area with no curbside recycling, she often does not even have the time to recycle. Elliott (1 child) reflected on how being a working mom with two jobs gave her more purchasing power to consume green but also limited the time she spent on sustainable habits: “I come into work at 9:15, just about, Monday through Thursday. Then, I also waitress two nights a week on James Island to just make ends meet. . . . If I wasn’t working, yeah sure, things become a whole lot easier. [But] then you don’t have money to sustain the lifestyle” (I-43, GP). Michelle (2 children) pointed to time constraints as the main reason that she still uses wipes, but that she tries to balance this out by being green in ways that are more accessible for her lifestyle:

> I still use wipes. . . . I’m just going crazy using wipes right now, but it’s . . . just . . . easier. . . . There are so many things that need to be wiped. . . . I think we make up for it in other ways. We don’t use plastic bags in the grocery store. I try really hard to always have a little, even if I go [clothing shopping], I try to have my own bag. A bag within my bag, to not use things. My kids see that, and they’re pretty good if we go to the grocery store [reminding me] “did you

> grab the bags?" . . . In our house we tried to switch out a lot of plastic . . . to the stainless steel or glass. I think the kids recognize that daily.[1] (I-13, GP)

Thirty-one percent of parents specified that it is not just that they were short on time, but also that they specifically lacked the investment time to learn a new routine. Rosy (2 children) discussed how "when you don't have [the money for] that fancy rain barrel or the fancy composter, you can do other stuff, but you have to have the research time, and then you have to have the time to do the extra labor that goes into making it work" (I-19, GP).

Similarly, Lindsey (1 child) explained that time to invest in a new habit would make sustainable living easier: "I think time is what everyone says [would make it easier]." Her husband, Pete, stepped into the conversation and added: "Money and time is usually everything" (I-11, GP). Then Lindsey added: "You can get around money if it's a priority, but it's time. Hopefully this summer, I'll finally be off because I've been in Montessori training in Texas the past two summers. This is finally a free summer where maybe I can dedicate that time to make it a habit" (I-11, GP).

This dual need for more overall time and investment time is related to some parents' frustrations, discussed below, with constraints imposed by green media.

MEDIA AS BOTH A RESOURCE AND A GREEN PARENTING CHALLENGE

Though I did not bring up the topic, parents revealed a complex relationship with media (including green media). Fifty percent of the parents did describe green media as a useful resource. Rose (1 child), for instance, explained her efforts in food preparation: "Well there is a learning curve. . . . [But] we're lucky. . . . You can just google it and quickly find 5 different ways to do things" (I-30, GP). Keenan (1 child), who heavily emphasized self-education through all types of media, put the same sentiment this way: "I'm doing as much research as I can. The internet is a wonderful thing, as much as you can sit there and waste all your time, . . . it is easy for me to look up these local farms and know who I'd like to buy from and what they do and realizing that organic doesn't always mean the best one around" (I-47, GP).

Rose and Keenan pointed to how media does involve navigation time but ultimately the value outweighs the investment.

Parents also voiced two main frustrations with media. Twenty-seven percent complained that green media was overwhelming and 19% noted that it was insufficient and inconsistent. Hank (2 children) reflected on his frustration with its lack of consistency: "I would appreciate some kind of audit that told me what . . . was really valuable. . . . [For instance], I should be doing a lot better at making compost . . . do I have to go out there with a pitchfork every two days and turn it? I guess I would just like to know what is the most beneficial" (I-1, GP).

He further explained: "I don't really know if [what I do] matters. . . . There are things that, even with recycling, the county operates at a loss because . . . no one wants it. You have to transport it, and the money it takes to transport it, you're not making it up. It can be frustrating to do the right thing, when you know it's not really making a difference" (I-1, GP).

Sharice (2 children) also shared: "I would like to know the most reputable blog or website that would tell me exactly what to do. . . . Really, just information, and not an overwhelming amount of options, but just like here's a good 12-step plan" (I-18, GP).

While Hank and Sharice experienced green media as insufficient, others perceived the constant onslaught of information as confusing noise, such as Jane (1 child) who shared her outrage about the controversies around Roundup pesticides:

> What starts to scare me, though, is when you see things like this Roundup stuff. This is stuff you think is safe 'cause they sell it in stores to kill your weeds. Then you read it's banned in Europe. . . . That really makes me mad when . . . you don't know what to believe. Is it propaganda? Is it the news? . . . You try to do the right thing and then you realize [you are using poison]. I feel like you gotta read so much. It can get so overwhelming. (I-23, GP)

Shelly (2 children) talked about her frustrations with navigating media and recalled that, when searching for information on eco-friendly family projects, she realized that she and her children had been painting seashells they had found at the beach, but "painting them with what? . . . Toxic, nontoxic, and then you can move into . . . painting them with Jello, and then, I'm like, 'Okay, but then red dye, huh?' There's just so much information out there" (I-4, GP).

NAVIGATING GREEN MEDIA AND CONSUMERISM

Practitioners' complicated relationship with media, especially green media, relates to parents' emphasis on being critical "conscious" decision makers (discussed in more depth in chapter 4). They actively resist the myriad of ways that green consumerist approaches to sustainable living moralize and individualize responsibility for environmental problems. For the parents in this study, they made efforts to avoid getting caught up in the dogma of being good green consumers. Instead of worrying about the latest green consumer trends, they focused on the holistic, pleasurable, and visceral experience of, say, being outside with their children. As one mother, Gaia (2 children), put it, green parenting is about the little things like lizards and bugs because "if you have respect and love for animals early on then you have that when you get older. . . . That is the cornerstone of what I believe, beyond buying the right products and things. . . . If you don't have that genuine love and appreciation, it doesn't matter what's [environmentally] trendy . . . if you're not feeling it" (I-12, GP).

Parents' conscious and critical navigation of green media and consumerism connects to another finding: while parents often expressed desire to do more, only 8% of the respondents mentioned guilt over not doing enough. Camelia's (1 child) story about why she ended up using disposable diapers is illustrative of how most parents pragmatically avoided guilt:

> I registered [for the baby shower] for some cloth diapers . . . [but] I was given a lot of . . . disposable diapers. [Since then], really every holiday, every time someone comes . . . over, it's a big box of disposable diapers. I don't wanna be ungrateful. . . . [So, I say] "thank you so much!" Then they think I want more. . . . I've got a whole garage full of . . . diapers. [Recently], I told [my husband], I'm like, "now that these poops are changing . . . I'm okay with cloth diapering." He was like, "well, does it make financial sense?" I said, "it used to . . . but in our case . . . it's really just extra money that we're gonna . . . spend and, sorry, it's just not gonna happen." (I-24, GP)

As I discuss below, interviewees were not just conscientious about consuming their own media, they were deliberate about monitoring their children's media consumption and saw this as part of their green parenting goals.

SCREEN TIME AS A GREEN PARENTING CHALLENGE

Even though I did not specifically ask about screen time, 31% of parents made a point of expressing concerns about their children's and household's time spent on phones, tablets, video games, televisions, and other screens. Interviewees, when asked what it means to be a green parent (GP), emphasized that it is important to de-emphasize screen time in the household. Parents expressed that they want to get their families outside and involved in other socially bonding and creative activities. One mother, Suzzie (1 child), summarized her approach to household media this way:

> We are very concerned about screen time and the impact it has on children, so we tried not to have any screen time at all until he was two. Now that he's two,

> he has limited screen time. If I don't have time to shower the night before, when he's asleep, sometimes I'll put him in a little travel crib and set up my computer and let him watch a PBS show. That's pretty much the extent of his screen time. . . . Instead we go outside to a park. I try not to be glued to my phone around my son 'cause I don't want him seeing that. (I-5, GP)

Another mother, Landon (3 children), complained:

> Video games and boys, it's a struggle in our house. I want them to get . . . outside. . . . For some reason they lost a little bit of that childhood wonder that I feel like I had when I was little. Like making forts. We don't watch that much TV [but] it's never easy. . . . I mean, just electronics in a world that's so driven by technology . . . sometimes the trivialness is overwhelming. "Oh my gosh. Have you seen this? Have you seen this?" What a waste. We just have this one life. (I-39, GP)

PRACTICING EVERYDAY FORMS OF RESISTANCE

I was initially surprised that so many interviewees talked about screen time when I asked about challenges to green parenting. In retrospect, and as I discuss in depth in chapter 4, parents were resisting media-fueled overconsumption and overwork by insisting on a holistic definition of green living that focused on well-being and not just traditional pro-environmental behaviors like calculating household footprints or purchasing the *correct* green products.

Parents' complex relationship with media, as both a resource and a stressor, is a reaction to the growth of the attention economy and the reality that twenty-first-century people (especially middle-class citizens in wealthy countries) are spending increasing amounts of time doing two things: working in paid employment and consuming media. A growing body of recent theoretical and empirical work examines how the digital economy of the twenty-first century is an attention economy in which consciousness itself is being commodified and treated as a resource (Crogan and Kinsley 2012; Doran 2017; Wu 2018, 2016). Peter Doran (2017), for instance, argues that the attention economy, with its focus on convenience and control (say, ordering groceries with Alexa), captures more and more of our attention and leaves us with less and less time for evaluation and reflection. Relatedly, a number of studies have pointed out that though modernism has created affluence for many, it has also created *time poverty* for the middle-class citizens of wealthy countries who increasingly live a life of constant work and compulsive consumeristic pursuit (Giurge et al. 2020; Humphrey 2009; Marglin and Schor 1990; Oriel and Gershuny 2001; Wajcman and Dodd 2017). Literary and popular press publications are also beginning to critique the overwork (Bunting 2004) and attention economies; Jenny Odell's (2019) book, *How to Do Nothing: Resisting the Attention Economy*, is one recent example.

Consumerist capitalism and the attention economy act in unison as oppressive social/cultural/political/economic regimes that demand participation and compliance, but that offer no clear focus for direct political action. With no reasonable institutional or political recourse available, parents told stories of actively resisting getting caught up in the attention economy and capitalist consumer culture in the place they exerted the most control—their daily lives. When they emphasized the importance of little acts, such as reducing screen time and making sure that their children got outside, they were engaging in what James Scott (1984) referred to as *everyday forms of resistance*.

CONCLUSION

These results echo the findings of other studies on the time challenges faced by working parents (Harvey and Mukhopadhyay 2007), especially green parents (Shirani 2013), and the multiple challenges faced by people living sustainable lifestyles. Newton and Meyer (2013), for instance, found that green practitioners named four key challenges to living sustainably: finances, time, lack of relevant information, and lack of understanding of how to get things done. In this book, I identified the same financial, time, and informational constraints among green parents. These analyses add to what is already known about the obstacles posed to green lifestyles by examining: how parents perceive and react to consumerism and, especially, green consumerism; and the important role of the attention economy in shaping the complex relationship parents have with media. The parents in this study, through their conscientious decision-making, and as I will discuss in chapter 4, through their insistence on defining "green living" holistically and across generations, echo *everyday forms of resistance* that are being practiced by many other pro-environmental practitioners across the globe. In the conclusion, I discuss how parents' challenges might inform policies and programs.

NOTES

1. Social practice theory would conceptualize Michelle's account of how, for example, she brings her own bags on shopping trips, but still uses disposable wipes, as *negative spillover*. Negative spillover occurs when participation in one pro-environmental behavior reduces propensity to engage in other pro-environmental ones and/or increases propensity to engage in wasteful behaviors (Nash et al. 2017; Nilsson et al. 2017; Thogersen and Olander 2003). With this understanding, Michelle

was able to justify, and thereby continue, her use of wipes by reminding herself that she was engaging in many other green behaviors. This study (GP) does not contradict the short-term spillover effect. However, it focuses on the efforts of green parents in a conscientious multi-generational holistic project that is taking place on a socially and politically contested terrain. Full success, therefore, cannot be measured in terms of today's constrained behaviors but only in terms of parents' *everyday resistance* (Scott 1984) that potentially could enable sustainability in the very long term.

Chapter 4

Green Living Motivations and Meanings

One of the most important questions of our day is, how do we encourage people to adopt green lifestyles? In order to begin answering this question, we need to understand what motivates green practitioners and what meanings they draw from sustainable living. Below, I discuss how parents are motivated by the socialization of their children, health, and thrift. Parents also held the following meanings: being conscientious, connecting to nature and to people, and well-being. Commuters were not asked about meanings, but they did reveal the following motivations: exercise and enjoyment, environmental concern, convenience, thrift, and self-sufficiency.

GREEN PARENTING MOTIVATIONS

For the forty-eight parents, four major themes emerged: the reduction of households' ecological footprint (forty-eight); the socialization of children to empower the next generation of green citizens (forty-eight); the protection of families' health, especially through diet (forty-four); and thrift (forty).[1] Parents (GP) articulated their motivations of reducing household footprint, living healthier lives, and being thrifty as everyday behaviors like keeping the lights turned off, eating unprocessed foods, and repurposing items, among many others. In the following quote, Hank (2 children) talked about his family's focus on health: "When you're a parent . . . you have more responsibilities, very serious ones, someone's well-being. . . . I don't want chemicals in their bodies. I don't want them going into puberty sooner than they should. . . . I try to be sensitive about shampoos or conditioners. In terms of food too . . . I buy certain meats, [especially] with no steroids" (I-1, GP).

Because they experienced challenges (culture, infrastructure, money, time, and media), parents innovated by focusing on their long-term motivation of socializing children into a greener future. In fact, children provided strong motivators for green living. Thirty-seven of forty-eight parents stated that, even though they faced challenges, green living was experienced as no more difficult after the birth of a child than before.[2] Only one parent reported that it was much harder with children and ten reported that it was a little harder. Many parents emphasized that, because they had already been practicing green living, it felt habitual and normal, and so they just didn't do anything different when the baby arrived. Callie (1 child) put it this way:

> We just kept doing what we were doing. I might've changed over to less, like trying to use apple cider [vinegar] for cleaning [to get the chemicals out of the house]. . . . It was already part of our lifestyle. We want it to be her lifestyle. We just didn't believe in the "we have a kid, so we have to change." We don't have to get a bigger house. We don't have to get a bigger car. We don't have to get all these toys. It's all *haves* that we just didn't wanna do [or] literally *buy into*. (I-32, GP)

Another explanation given for how and why the transition into green parenting was so seamless is that while children demanded more resources, they provided an even larger reservoir of motivation for parents to live pro-environmental lifestyles and to pass those lifestyles on to their children.[3]

This deliberate long-term focus on the socialization of children into green living emerged in the narratives of Roselyn (1 child), Sylvia (1 child), Germaine (3 children), and Eliot (1 child):

> For me, it's knowing that I'm making conscious efforts to make the world cleaner, and safer, and still a wonderful place for her future. . . . That, to me, is green parenting. My efforts will teach her how to start now to lead a [green] life. . . . We're setting the foundation for a better, greener, world ahead, and raising children that are environmentalists from day one. (I-17, GP)

> I have some faith that it is making a difference. I want my child, and her children, to grow up in a world that's cleaner and greener and not full of as many landfills. I want her to be able to eat fish and animals that are not full of plastic because I've contributed—I only have control over what I have control over. . . . Every little thing has a broader impact that builds on itself. (I-28, GP)

> What you instill in your children is gonna last generations 'cause they're gonna do it with their children and so on. . . . They're gonna be greener folks and know how to grow [food] and teach their kids how to do this because of what they were exposed to from a young age. (I-48, GP)

> I want my daughter to understand consequences of her actions environmentally. We don't use half a roll of toilet paper to wipe our butt because . . . you're

> polluting. We don't use plastic water bottles because they're not environmentally conscious. We don't use paper towels to wipe our faces. We use hand towels. I think as a parent, these are very, very, very small decisions. They're things you can implement with your children from day one that just make them a little bit more aware. I hope that by implementing that, and making it what she is used to, then, hopefully, she'll continue living that way. That's such a small change, but I feel like, if everyone can make such small changes, that globally it would have a significant impact. (I-43, GP)

In their implementation of this long-term innovative strategy of socialization, parents emphasized holistic and interconnected meanings. Rebecca (1 child), when discussing the importance of connecting her daughter to nature, told a story about food gardening:

> Last year I did it for the first time [with my] daughter who was then three. We got cantaloupe and I was like "where do cantaloupes come from?" She was like "a plant." I'm like "yes!" . . . Now we get up every morning, we look at the flowers, and now that the cucumber's coming out, the flower is at the end. She sees how things grow. She's like, "Those look like cactuses." I'm like, "yeah, and they're pokey like cactuses too." (I-36, GP)

While Rebecca focused on connectedness, Michelle (2 children) and Antoinette (1 child) discussed passing on green consciousness:

> My son's doing a weather study right now, so he's . . . learning [that] this is the earth and . . . how can we take care of it better? That's a big conversation in our house these days.
>
> . . . [I want my] kids to grow up in a world that is conscious about the world around them and to want to protect the earth so it's around forever . . . and not pollute the air and the water and everything. I hope they'll be able to enjoy it and . . . their kids and their kids' kids. (I-13, GP)

> I guess it's making sure that I'm aware and making my daughter aware of the resources that we're using, whether that's water or electricity or gas or food or stuff, toys, and clothes. . . . Just having that awareness and realizing that, when you're making decisions, you are making decisions to be sustainable or not, whether you're thinking about it or not. Just having that be a part of the conversation every day, I guess. (I-26, GP)

My analyses extend Fiona Shirani and colleagues' (2016) findings, from their study on ecovillages, that children can be a sizable motivator for parents who choose to live sustainably. While the Shirani et al. study focused on highly dedicated green practitioners, my analyses show that, even for everyday people, children can, and do, act as motivators for green living.[4]

Research demonstrates that parents are correct to focus their energy on pro-environmental socialization, especially talking about the environment, modeling green behavior, and getting their children off their screens and outside in nature. Studies consistently show that when children and adolescents are exposed to pro-environmental behavior and dialogue it has both immediate (Barrera-Hernandez et al. 2020; Collins 2015; Gronhoj and Thogersen 2012) and long-term positive impacts (Armstrong and Impara 1991; Bakir-Demir 2019; Barton 2016; Charles et al. 2018; Chawla 1999, 2001, 2008, 2020, 2021; Chawla and Flanders 2007; D'Amore 2016; Gronhoj and Togersen 2012; Stevenson et al. 2016; Williams and Chawla 2016) on sustainable beliefs/behaviors. Stevenson and colleagues (2016) found, for instance, that the frequency of discussion about climate change with family and friends was a strong predictor of adolescents' climate change concern.

Several studies document a positive connection between getting children connected to nature and the maintenance of pro-environmental attitudes and behaviors, many years later, in young and older adulthood (Broom 2017; Duron-Ramos et al. 2020; Krepelkova et al. 2020; Molinario et al. 2020; Rosa et al. 2018; Wells and Lekies 2006, 2012; Williams and Chawla 2014). One recent study, for example, documented that childhood nature experiences were positively correlated with adult recreational time in nature and that adults' nature recreation was positively associated with their connectedness to nature and pro-environmental behaviors (Rosa et al. 2018). Another recent study found that, over a twelve-year period, youth who grew up with mothers with pro-environmental attitudes and, also, youth who grew up spending a lot of time outdoors engaged in more pro-environmental behaviors as young adults (Evans et al. 2018).

Studies also show that programs aimed at encouraging young people to connect to nature can positively impact youths' overall well-being (Chawla 2021; Collado and Corralza 2015; Dopko et al. 2019; Harvey et al. 2020; Hatala et al. 2020; Kuo et al. 2019; Pirchio 2020; Tillmann et al. 2018; Windhorst and Williams 2015) and adult well-being later in life (Wood and Smyth 2020). A recent study found that schoolchildren's participation in a biodiversity program resulted in significant improvements in their mood (Harvey et al. 2020). Performance in school has also been found to be positively correlated with connectedness to nature (Kuo et al. 2019; Musitu-Ferrer et al. 2019). Kuo and colleagues (2019), for instance, documented that exposure to nature promotes children's perseverance, problem solving, critical thinking, leadership, and teamwork. Louise Chawla, a longtime expert on education, children, and nature experiences (2021),[5] conducted a comprehensive review of the social science research, theory, and philosophy on this topic and confirmed that the weight of the literature shows that outdoor experiences present opportunities for young people to express agency and self-determination in natural environments. In this way, she concludes, connections to nature can help children *fully realize their lives*.[6]

In addition, research indicates that parents are correct to be concerned about the relationship between outside time and screen time. Studies are beginning to unravel a troubling set of interacting relationships between increased youth screen time, decreased outdoor time, lower feelings of connection to nature, and lower developmental skills (Hinkley et al. 2018; Larson et al. 2019; Mutz et al. 2018).[7] Trina Hinkley and colleagues (2018), for instance, found that, for preschool children, the more time they spent on screens, and the less time they spent outside, the more poorly they scored on social skills. Lincoln Larson and colleagues (2019) documented a relationship between increases in screen time and declines in both outdoor time and feelings of connection to nature among youth.

Lowcountry green parents modeled and taught green behaviors/attitudes, fostered a love and respect for nature, and emphasized outside time for their children. These efforts were informed by several interlocking meanings that I discuss below.

Green Parenting Meanings

Parents articulated the following four meanings of green living: being conscientious decision makers (85%), connecting to nature (81%), connecting to people (67%), and cultivating a sense of well-being (71%).[8]

Conscientious Decision Making

Conscientiousness was expressed as deliberate decision-making within constraints. In Ari's (3 children) comments below you see how, while she was not engaging 100% in what she felt was green living, due to finances and inadequate public transit infrastructure, she was thinking carefully about the impacts of her choices. In her words:

> I like to buy in bulk. I try to buy things that use less packaging. I would never use [single- use] coffee [pods]. . . . We use a French press. We try to use less plastics. [Our] kids drink out of metal cups. We had to buy a minivan, but we would have bought a hybrid or any type of electric car if we could afford it. That's really what we wanna do. I lived in Chicago for so many years, and I didn't even own a car for a long time. We are 100% about public transportation. (I-3, GP)

While Ari expressed the desire to be able to afford to buy a hybrid or electric car, Seneca (2 children) discussed how she and her husband deliberately did not buy a hybrid car: "We got a sedan, just a car, because it's better on gas. We didn't get the eco-car [even though] it has really good gas mileage [because the batteries are toxic]. . . . [It was] a more reasonable choice than an SUV or some big truck . . . for us. Just buying a car that's a more fuel-efficient car than the one you had before is a better choice" (I-25, GP). Cab (3 children) referenced green media when she talked about how you have to be pragmatic about all of your green choices. She explained that when she sees "all those blogs of natural things you could do [like making your own detergent] . . ." she asks herself if it will be efficient and if it is actually going to be cheaper and make an ecological difference. Often she ends up saying "that's one step too far. I'm not really sure that it would get my clothes clean. I don't know if I wanna try it" (I-2, GP).

As part of this deliberate decision making, parents balanced their own and their family's overall psychological health with attempts to leave a smaller footprint. Lucy (2 children) talked about how she feels like if she were "extremely careful about every specific decision all of the time, I'd lose my mind. There'd be too much for me to handle worrying about every little thing and trying to control every little thing. I have to let some things go. So, I try to keep a balance" (I-7, GP). Sharice (2 children) discussed how her family emphasized eating whole rather than processed foods but that, sometimes, there was not enough time to cook and to shop. She explained: "I try to always remember that stress is bad for you too and if it's between the two, I'll go for the quick food" (I-18, GP).

Some parents defined consciousness, not only as deliberate decision making, but also as critiquing green consumerism. Anne (2 children) made the following observation: "I think being green covers a lot more than the advertisements or the buzz that people put behind it. I think it is even just being conscious about your purchases, or not purchasing, just because it's extra stuff" (I-22, GP). Likewise, four interviewees spontaneously brought up that they explored solar panel installation for their homes but decided against it because (due to tree shade and other factors) it didn't bring environmental benefits to their household energy use. Suzie (1 child) recollected how she and her husband researched solar panels. A company came to her house to give an estimate and

> . . . We went outside, and we're looking at the house. He's going, "(Y)eah, so you have this big, huge, tree here that's overshadowing your house. You'd wanna get rid of that." I was like, a tree is protection from storms and it is natural heating and cooling, natural heating and cooling effect right there! It's a

> tree taking water from the ground, flood control, the roots are keeping the soil in place. I was like, "no!" (I-5, GP)

This critical approach to all types of consumerism connects to another finding, that while parents often expressed a desire to do more, only 8% mentioned guilt over not doing enough. I asked parents about challenges, not guilt, but I noticed an absence of talk about guilt in the interviews. A few of the interviewees were explicit that guilt was not an appropriate way to respond to areas in their life where they were not living 100% green. Isabelle (1 child) made it clear that it is important to not feel guilty, or allow others to feel guilty, about what they are not able to accomplish. She talked about how she helps people by

> being someone that other people can just see and just be a person who does it. Also, maybe, just being there, especially for people who . . . may be really busy and just being able to say, "yes, I can't do everything I wanna do, but every little bit helps." You kinda have to pick and choose. I may not be able to do everything I could be doing or everything I should be doing, but I can do these little things. At least that helps things, anyway. (I-21, GP)

Instead of talking in terms of guilt, most parents talked about how there are trade-offs and that the important thing to do is to find a balance. Hank's (2 children) family invested a sizable amount of time and energy in green living, including: installing geothermal heating in their house, organic gardening and composting, researching toxins in products, purchasing a hybrid car, and actively educating their children about respecting nature and animals, among other activities. When asked about the challenges of green parenting, Hank talked at great length to explain how sustainable living involves trade-offs at the household level and also at a societal level:

> I guess the first thing was diapers that came to mind. I have two children. . . . When (our first daughter) was born, we had bought a number of cloth diapers. . . . They really do take more time to clean. They are not as effective . . . they leak more, they are messier. We did it probably for six months or so, and then we just slowly gave it up. Travel was difficult, especially. . . . Another concern was just it's not sanitary. We're spreading feces. . . . Really, health reasons, sanitation, water usage, and then just the inefficiency of some of the cloth diapers convinced us really to slowly change to disposable diapers. . . . I've tried to find a happy medium, a balance. . . . (I-1, GP)

He went on to talk about how, for his family, making environmental decisions involves trade-offs between spending time with their daughters, running a dual-income household, and working at a career and a side gig as a journalist.

Part of the trade-off, he explained, involves maintaining the necessary social status that is required by your career and other areas of life: "I will buy a sweater every two or three years instead of wearing one that's ten years old. . . . There's a standard of living that requires a certain amount of energy, and I'm part of that. . . . We all know, what certain people feel when they see a hippie. I don't care if they're a hippie. . . . But some people may not be promoted as quickly, or at all, if they're seen as different or not fitting in" (I-1, GP). He connected his personal decision making to larger societal trade-offs and explained:

> It's a hard thing, I think, just recognizing that in civilization, there are gonna be negative impacts on the environment. If we're gonna pave roads, that means taking a pristine environment and covering it with asphalt. If you're gonna be making certain things that are critical, say, medicine . . . there's gonna be some kind of waste. . . . We'll be overrun by other armies if we don't keep up with our military. . . . That's the way it is. Some of us are very sensitive to it, but you have to realize there are impacts to the way we're living and choosing to make our lives better. . . . Striking a balance is the challenging part. (I-1, GP)

It is not just that parents did not express guilt, many also made sure to explain that, while they were concerned about the environment, their higher priority was healthy social connections. For instance, Avery (2 children) told a cautionary story about a friend who, she thought, got his priorities out of balance:

> I think mostly it's about figuring out how to live sustainably within our culture. . . . I have a friend who didn't fly for a few years . . . to cut down his carbon footprint. Then he was supposed to be in a wedding. He didn't go because he would not fly.

She concluded: "For me, I don't see any reason to have a better relationship with nature if it's getting in the way of my relationships with other people" (I-35, GP). Avery, like other parents in this study, was conscientious about living green while not getting caught up in the moral messages that come from rigid conceptualizations of environmentalism.

Another important meaning expressed by parents was connecting to nature and people.

Connecting to Nature and Each Other

Eighty-one percent of parents brought up the importance of connecting to nature, especially connecting their children to nature. Mandy (2 children), Sylvia (1 child), and Callie (1 child) expressed it this way:

> The kids are outside all the time, especially my son. . . . They just love it. We like goin' and bein' outside. Since I stopped working, I started walking and bicycling. In the mornings I walk, and then, in the afternoon when my son gets home, we usually go for a bike ride. (I-33, GP)

> We're fortunate to have a big backyard with lots of shade. Whether it's just playing out there or going to a park, I try to let her get dirty and pick up bugs and find stuff. (I-28, GP)

> She loves bein' outside. She's so observant. She'll [notice] birds flying. She gets so excited over other dogs. She met a goat a few weeks back and was so excited over that. Yeah, she has that excitement. I'm excited for her. (I-32, GP).

These activities were part of a long-term project to instill a love of nature in their children. I talked to Abby (1 child) while we sat in her backyard. She looked over at her daughter pushing a toy through the grass and commented:

> I want her to . . . enjoy the outside a lot more. . . . I show her how to recycle. We reuse a lot, too. If we use a soymilk carton, sometimes we'll make a bird feeder out of it and put it in the backyard. Then, when it falls, I'll recycle it. She planted a little flower right there. . . . I just want her to be connected with this Earth more than many kids are, because they're so connected to the internet right now. Screen time. There's no outside time for them. If kids don't look outside and enjoy the outside, why're they gonna care about the environment? (I-10, GP)

Respondents often talked about connections to nature and people in unison. Muir (1 child) lamented:

> We're losing our connection with nature . . . 'cause we're in an environmental crisis. We've lost our connections to our neighbors 'cause we're in a social crisis. Since having [my daughter], I believe it's even more important to pass those values down to her instead of it being lost, which is seen so much today. . . . If you have more of a relation or connection to your surroundings, you're more likely to protect and respect it. (I-20, GP)

Stella (1 child) went further to explain that it is not just about feeling connected, but she also wants to leverage her connections to build up her natural and human communities, to "support your neighbors and be a neighbor. . . . I love making community connections. 'You have a need. I know this person that can fill that need. Let me connect you all so that we can create just a really strong, vibrant place to live and help everyone thrive.' That to me is really important" (I-40, GP).

Interviewees experienced their love of nature exactly because of strong connections to their communities and to the most important people in their

Figure 4.1. A low country child playing outside and repurposing a box into a fort before it gets recycled. ***Photo by author***

lives. As I show in more detail below, this is part of parents' holistic understanding of green living in which pro-environmental behavior represents only one layer of a lifelong project of cultivating well-being for both human and ecological communities.

Well-being and Living a Meaningful Life

One narrative that weaves through many (71%) of the interviews is the sheer joy and visceral pleasure of everyday green activities, such as cooking a healthy meal or inspecting bugs on a nature walk. I call this the desire for *well-being and wanting to live a meaningful life*. Sharice (2 children), Kareena (2 children), and Keenan (1 child) put it this way:

> I think [green living] has to involve more than one area of life, so in other words, it's not just recycling things. I think it has to involve your well-being. (I-18, GP)

> I guess the thing that comes to mind is just the joy that I have with my daughters when we go to the park, and lay out on a blanket, and open up some picture books and start reading on a pretty day. I want that to always be an option for everybody. I know that's kind of general, but I feel really strongly about . . . not taking the beauty that surrounds us for granted. And sharing that part of it with my daughters is my number one, hands down, the number one best thing in life. I don't need to go to Hawaii, I don't need to travel or get the coolest new gizmo or whatever. I'm happiest to be just outside with my daughters, enjoying it. (I-29, GP)

> The environment, what goes in our bodies, and what I put out in the world. I'm at a point in my life where I feel completely tied to what's around me. What's around me is tied to me and we all affect each other. Whether it's people, or a place, or an ecosystem, or nature, or animals. I'm just going to do what's best for it all and it will do what's best for me. (I-47, GP)

Avery (2 children) recollected how she learned about environmental degradation in college and started to make a commitment to live that way. However, it was later, when she moved to the mountains and experienced the joys of the outdoors for the first time, that she became serious about living a green lifestyle. She went on to explain that a big part of what it means to live green is "just enjoyment. [And] I think part of it is resisting consumerism. . . . I think to slow down, to not be going all the time, or overschedule. . . . It all goes back to the bigger picture" (I-35, GP).

As I noted earlier in this section, parents faced challenges and sometimes failed to live as green as they wanted, but most did not express guilt over not doing enough. I was surprised to find that so few parents expressed guilt, as several studies link the individualization of environmental responsibility, like green consumerism, to the experience of guilt over not living *green enough* (Elliott 2013; Fredericks 2014; Huddart Kennedy and Givens 2019; Jayaratane et al. 2015; Karlsson 2013; Mkono 2020). As I discuss throughout this chapter and much of the book, parents see their lifestyles as part of a larger project of socializing their children into sustainable living, being conscientious decision makers, prioritizing connection to nature and each other, and seeking well-being. Their deliberate efforts to make conscientious decisions appears to serve as a bit of protection against the misplaced guilt that can be a result of green consumerist approaches to sustainability. Their focus on their children's futures and their prioritization of well-being gives them a healthy perspective on where to spend their energy. They are clearly saying, *green guilt is a waste of time* and they aren't *buying into it*.

While green parenting represents an intentional lifestyle, cycling represents one way to engage, sometimes intentionally and sometimes unintentionally, in a larger project of green living.

CHOICE COMMUTER MOTIVATIONS

I asked parents about both why they engage in sustainable living (GP) and what it means to them. For cyclists, I only asked about why they commute (CB or NB). The forty-one choice cyclists identified the following multiple and overlapping motivations for currently commuting: to get exercise (thirty-three); out of environmental concern (thirty); for enjoyment (twenty-six); for thrift (twenty-six); for convenience (twenty-five); to connect to nature (nineteen); to connect to people (thirteen); and to feel self-sufficient (five).[9]

Most cyclists named exercise and enjoyment as their primary motivators for commuting. In this way, they echoed parents' emphasis on well-being as an important meaning of green living.

Exercise and Enjoyment

The high numbers of respondents (80%) who named exercise as a strong motivator is not surprising given that 76% pointed to how their commuting originally evolved out of exercise and/or recreational biking that they were already doing. Exercise and enjoyment (63%) were often brought up in unison. Vince described his exercise-driven bike commute this way:

> My first couple of pedal strokes are stiff and then, after a mile, I start to loosen up, and everything starts to feel better, and I start turning on the heat, and by the time I get to the bridge to go up the hill I'm loosened up, and that's a challenge no matter what. It's going to make you hurt a little bit. And I try to do a little faster each day for my training. Coast down the back side, by the time I get on to the other side of the bridge, I'm feeling good. You know, moving fast on the bike, weaving through neighborhoods, beating traffic which is stuck waiting to make that right-hand turn. Then I make it a game. I'm going through downtown

> doing sprints and stuff. So now I'm at work. I'm wide awake. I got a healthy bead of sweat going. I towel off. . . . I don't need coffee. . . . I'm more relaxed throughout the day. (I-6, CB)

While some riders, like Vince, desired an intense workout, others focused on the practical need for exercise as the primary motivation. Howard noted: "I am not the type of person who likes exercise. I never run. I just don't have any enjoyment in it. So, the fact that I can kill two birds with one stone by getting to work and getting exercise, it's a lot more sensible to me" (I-25, CB).

In terms of enjoyment (63%), many of their stories focused on the pleasures of being a moving body travelling through and interacting with nature. The words *free*, *fun*, *energizing*, *Zen*, *groove*, *peaceful*, and *liberating* appeared throughout their narratives. Some explained this joy in physical terms. Hudson, a young woman who could afford a car but chose not to own a car because she doesn't like driving, declared simply: "You know how it feels when you taste something for the first time and you just want to scream because it is so good? It's how I feel about biking" (I-4, CB).

Others described biking in meditative terms. For Fran: "There's a kind of Zenism, a kind of state of consciousness where you can . . . let your . . . mind wander. . . . I find it's really stress relieving, mind focusing" (I-1, CB). Mark also talked about how he feels more aware on his bike. Where Fran talked about this in peaceful terms, Mark talked about how the exhilaration of the commute heightened his senses:

> The sights and smells are so poignant when you're on a bicycle. . . . Four miles for me to go from here to [law] school and I go through very different areas. First of all, down the greenway. . . . Then I cross over some hellish traffic and even that's kind of fun because there is an element of danger or high-speed chase. You know, you are dodging cars. . . . It's kind of fun just sort of always being really aware all the time. (I-2, CB)

Several, like Jenny, described biking as freeing:

> Oh, there are so many things. It's liberating. . . . I feel like I'm free. I'm never stuck in traffic. . . . And, you know, you've got the wind in your hair kind of thing. It's just a great feeling. And, you know, exercise really helps make people deal with any kind of stress. . . . You just need a release. . . . Riding my bike home is wonderful because it's just kind of like the end of the day, kind of let it all go. . . . I think I have the most beautiful commute in America, with a ride on the Ravenel Bridge. (I-15, CB)

Caleb also captured this freeing feeling when he compared bike commuting to a pelican who "doesn't really fight the wind. It just kind of, I don't know, glides. . . . Sometimes you just get in the groove" (I-19, CB).

Forty-four percent of the respondents also told less-charming stories about, for instance, feeling grit hit their faces and smelling the exhaust after a large truck passes by (I-18, CB). But, on the whole, even those experiences were recounted in ways that made it clear that any sensory experience on the bike was better than the cheerless void of driving in a car. Todd talked about how he was "enthralled by everything that I see on the road that I would not see in my car. I mean, roadkill, or animals in salt marshes, getting good looks at the birds that I would just zip by at 60 miles per hour" (I-23, CB).

Environmental Concern and Connecting to Nature

Over half (73%) of choice bikers indicated that environmental concerns currently motivated them to commute. This is surprising in that only 32% listed such concerns as a pathway to commuting. Chuck, for instance, talked about how he started bike commuting for the exercise, stress release, and self-sufficiency, but that the reason he does it now is also about the environment: "Part of it is wanting to divorce yourself from dependency on a car. I think part of it is that getting to work and getting home with a clean, fresh happy mind and happy heart is really rejuvenating . . . and part of it was just fittin' in exercise first thing in the morning because if I don't get it there, I won't do it" (I-14, CB).

He went on to explain that, while all these factors explain his becoming a commuter, over time his biggest priority became "the impact that vehicles have on pollution" (I-14, CB). Colleen, likewise, talked about how she feels good that her biking benefits the environment but that she originally started commuting to work to avoid parking costs. In her words: "I liked that it was [reduced] parking costs, and then later the environmental [benefits] made me feel better about it." She went on to explain that she is now planning to take the bus [instead of driving her car] on bad weather days because the environmental importance of biking has grown for her (I-9, CB).

Howard put his environmental interests this way: "I'm glad that I ride my bike. There's an intrinsic value knowing that it's not just a cost saving thing, but . . . if I ride 600 miles in a year, I'll save x number of pounds of carbon" (I-25, CP).

Luca discussed the importance of his environmental concerns in more detail:

> One of the things I tell myself as I'm riding in traffic, if I'm feeling intimidated or if I'm feeling like I'm imposing on the drivers, I remind myself that, yes, my footprint is a lot lighter on the earth. And, as such, I deserve my space because

> it's very, very little space that I require. . . . It fits in with the whole idea of . . . don't leave footprints behind. You know, when you pack it in, you pack it out. . . . My daddy always told me that whenever he went into a store . . . he had an early career in retail, "I always try to leave the display better than I found it." And that kinda stuck with me. And part of what I try to do, and part of what bicycle commuting does is, I don't disturb my environment as I move through it. . . . I'm not interested in leaving marks of where I've been. (I-39, CB)

Related to environmental concern is respondents' desire to connect to nature. While none of the choice bikers indicated that a desire to connect to nature was a pathway to commuting, almost half (46%) talked about how it represented a sizable current motivation. Cyclists cited sounds, smells, touch, and sights to talk about the joy of connecting to nature on a commute. They named the smell of pluff mud, the sight of dolphins swimming in the Ashley River, the feel of transitioning from pavement to dirt to grass, and the sound of gulls fishing, for instance.

Jim described "the beauty of getting up on the bridge and looking out over the harbor, seeing the play of light on the ships . . . hearing the birds calling and clapping of the rails . . . looking down and smelling the marsh at Shem Creek. If you're in a car, you don't get to do that" (I-18, CB).

Often when respondents talked, they drew on a broad understanding of nature as existing in both the natural world and the built environment. This comes through in the following quote from Henry. When asked why he continues to bike commute, he gave that day's bike ride as an example:

> I set out to the barber, I looked up and saw some kind of threatening clouds, and I thought, hmm, I hope I'm home before that cloud bursts. . . . The commute to the barber is five blocks. . . . Just going up East Bay Street and taking in the license plates and the work going on in a building sort of not far from where I live—wondering whether that's a new restaurant or what's going on in there. . . . I was open to all those things. Whereas if I was driving, I think . . . I might have seen some of those things but not as freely as I did. . . . [Charleston] is about the water, it's about the historic nature of the area itself, the buildings and homes, it's about the marshes, it's about the beaches, it's about all those things that make people want to be outside. And the great tragedy is that [people feel like they have to] [contain themselves] in a steel enclosure to get to the outdoor pursuits. (I-3, CB)

In addition to the pleasures of connecting to nature, cyclists pointed to the importance of connecting to people while on a bike.

Connecting to People

Thirty-two percent of choice cyclists listed connecting to people as an important motivator for cycling. For some, that meant simply communing with people in a shared space. Hudson, for instance, talked about how she enjoys biking by the water view along the Battery and seeing the people, "because there are so many different personalities . . . little kids out there playing on the playground. And so, there is just so much to see and enjoy" (I-4, CB). Others emphasized how much they valued interactions with people. Arnie mused about how he enjoys

> the community. I was at an intersection on Calhoun and Rutledge, and a guy comes up beside me and asked for directions and, if I was in a car, he wouldn't do that. . . . I think there is a lot that people miss being in a car. . . . It's like intangible, like if I meet another biker around the corner or at the bike rack, or I am loading my groceries at the bike stand, I think I talk to more people that way than if I am loading the car, closing the doors, and rolling up the windows, and turning on the radio. . . . Cars make people exclusive and I think bicycle riding is more inclusive. (I-5, CB)

Likewise, Jason talked about how he likes that there is a woman who works near where he lives and they pass each other and say hi from their bikes (I-30, CB). Louise talked about how she often speaks to people while commuting: "Sometimes people are out in their yard and you say good morning. . . . There's more of a connection. It feels much more like a community. Where you know people, and you don't necessarily know their names, but you might see them all the time. And other people commuting. It's sort of a nice way to wake up, you know, you are just out there" (I-38, CB).

Others talked about how they like seeing their friends. Ted recounted about one of the first times he commuted in Charleston: "I'm going across the bridge and I pass an old Sierra Club buddy on his bike. I hadn't seen him in forever, so it was great and we're chatting. The next Friday that I rode, I saw [another friend] sitting outside at a little shop, eating breakfast, and I stop and talk with her. The next week, I saw [yet another friend] on his bike" (I-20, CB).

Choice bikers also named the pragmatic benefits of thrift and of convenience as motivations for commuting.

Thrift and Convenience

While only 15% of choice commuters indicated that financial necessity was a pathway to commuting, over half (63%) pointed to thrift as a current motivator. At the time of the study, everyone could afford to afford to purchase and maintain a car, and most of them did. However, being conservative with money is still an important motivation for commuting. I imagine that would be even more true today, in 2022, with our historically high gas prices following Russia's invasion of Ukraine. Katarina put it this way: "I find myself gassing up my car and it's thirty-something dollars [chuckling] a pop. That's thirty dollars a week . . . that's significant" (I-8, CB). Vince, likewise, explained:

> I could afford to drive a big Lexus SUV, I drive a little Subaru Outback Sport. But a waste of money is still a waste of money. But you have to talk beyond gas when you talk about this. I just bought a $1,500 bike, a very nice bike. . . . I very much enjoy riding it. . . . At $3 a gallon, I'm saving a few bills there. But I'm also not putting miles on my car, so that all adds up. Plus, I don't have to buy a new car so there is a lot of value added that way. And the bike doesn't require insurance. (I-6, CB)

Though none of the choice bikers brought up convenience as an original path to commuting, over half (61%) pointed to it as an important current motivator. Evan and Jimmy described convenience this way:

> You can get anywhere [on the Charleston peninsula] on a bike. It's faster. There's nothing better than riding by a traffic jam. It's a good feeling being able to get around a lot better than cars. (I-36, CB)

> I live a mile and a half from work. . . . A straight shot down Wentworth Street. . . . I find it easier to negotiate traffic in a bicycle. . . . Less frustrating, less irritating. In an automobile, as we all know, you often get caught in heavy traffic and there's the road rage factor . . . especially around rush hour in downtown Charleston. Those are the negatives for driving. Whereas on the bicycle, one can often change a route and take backstreets or go up a sidewalk, if you need to, to get around obstructions. . . . I find it very convenient to . . . go right up to the door and station the bike, lock the bike there, and walk in, rather than looking all over for a parking place. (I-24, CB)

Thriftiness and convenience were often brought up together as co-beneficial motivators. When asked for one word that would sum up bike commuting, Tony offered "simple" and explained:

> The most rewarding thing about bike commuting is that, one, it's predictable. I always knew how long it was gonna take to get to work. There's not a lot you can do to slow me down. . . . I don't get into traffic jams. . . . I'm not affected by transit strikes. I ride my bike. . . . It's nice I don't have to worry about parking a car. I don't have to pay for that. It's nice I don't care what the hell gas prices

are. . . . I mean, financially, I could afford the gas, could afford the car. . . . It just really doesn't take that much effort to get somewhere on your bike. (I-37, CB)

Self-Sufficiency

Twelve percent of choice bikers pointed to self-sufficiency as an important motivator to commute. Ginny talked about how she likes "relying on nothing but you" (I-7, CB). Luca, likewise, explained: "I know my bike like the back of my hand. I put it together, I take it apart, I fix it, I replace all the stuff on it. If something goes wrong on my bike on a commute, it's not a showstopper. I don't have to call anybody else. I can make it work again. . . . A bike is almost an extension of myself. My muscles and the bike are working together" (I-39, CB). While only 12% of the choice and none of the necessity interviewees talked directly about self-sufficiency, the desire to master the bike and avoid dependence on a car were woven, in indirect ways, throughout many their narratives.

Below I discuss necessity commuters who shared many of the same motivations as choice commuters.

NECESSITY COMMUTER MOTIVATIONS

In chapter 2, I explored how necessity cyclists (NB) started commuting—they couldn't afford a car and/or lost their license; they had a history of riding and they wanted exercise; and they wanted the convenience. The themes of exercise and convenience showed up again as current motivators, along with enjoyment, thrift, and connection to nature.

None of the necessity commuters could afford to drive or maintain a car but they did not bring up financial necessity as a current motivator for commuting, though this was certainly implied. What they did emphasize in their narratives were more leisurely motivations like exercise and enjoyment.

As with parents and choice cyclists, necessity bikers listed multiple and overlapping motivations for commuting. Louis put it this way: "Biking is good exercise and I get to see a lot [and] go places I want to go when I want to go" (I-11, NB).

The twenty-one necessity cyclists pointed to the following motivators: enjoyment (thirteen), exercise (nine), thrift (five), connection to nature (two), environmental issues (one), and convenience (one). The most striking similarity between the parents and the choice and necessity cyclists was their emphasis on enjoyment.

Enjoyment and Exercise

Sixty-one percent of necessity bikers emphatically voiced enjoyment as a motivator for commuting. For instance, Dean, Reginald, and Thomas exclaimed:

> Anybody say bicycling ain't fun, they got to be crazy. (I-15, NB)

> It's just fun, I just love riding my bike. (I-14, NB)

> I love bikes. I love bikes. . . . That's what I like to do, especially when I feel angry, I can go down Bull Street by the water. They have a pond down there by Ashley Avenue and I'll leave my bike by the side and get in the water. (I-17, NB)

Forty-three percent pointed to the desire for exercise as a motivator. For instance, Charles explained that "riding a bike is healthy for you, for you will lose weight, exercise more on the bike you use to get around town. It's the best thing, I think, for people old or young" (I-21, NB).

Thrift and Convenience

Twenty-four percent brought up cost savings and 0.05% brought up convenience. Ernie and Reginald put it this way:

> To me, you got gas [to pay for], then you get in an accident more quicker [in a car]. With a bicycle, you avoid that. You ain't got to worry about no gas or nothing. . . . You ain't got to spend no money, only thing you spend, only money you spend, you get a flat tire. That's the only money you spend, you hear. (I-4, NB)

> It's nice because you're exercising, but it's still getting you around. It's actually faster, most of the time, than driving a car, especially downtown with traffic and everything. It's a lot quicker. It's a lot more convenient. You can park it anywhere, that kind of thing. (I-14, NB)

Environmental Concern and Connection to Nature

One necessity cyclist, 0.05%, indicated that environmental concern was a current motivator for commuting. Alejandro explained, "It's healthy. . . . That's basically, you know, what a bicycle is all about. Cuts down on pollution and transportation" (I-1, NB).

The stories of parents and cyclists point to motivations and to meanings that expand way beyond pro-environmental behavior. Below, I discuss how practice, innovation, and embodiment research shed light on these shared understandings.

CONCLUSION

In the introduction, I established how, while everyday individual behavior change is not sufficient on its own, it is a potentially important component of larger-scale successful environmental change. Yet, how do we go about enacting individual and household-level change? The dominant paradigm, which asserted that providing information or raising awareness would lead directly to changes in behavior, has been challenged by decades of research (Barr 2006; Blake 1999; Chawla and Cushing 2007; Heimlich and Ardoin 2008; Hines et al. 1987; Jensen 2002; Kollmus and Agyeman 2002; Shove 2010; Siegel et al. 2018; Stern 2000).

Practice theory offers an alternative paradigm for understanding how individual and household-level change takes place (fully cited in the introduction). For this body of theory and research, practices do not arise directly out of individual beliefs nor out of wide-scale policies but, rather, at the meso-level from ongoing interactions between individuals and social structures. It thus offers an excellent framework for understanding how both parents and utility cyclists carry out green lifestyles.

Practice research illustrates how people draw on a mix of meanings, resources, and skills to enact their lifestyles (Hargreaves 2016; Lorenzen 2012; Reckwitz 2002; Shove et al. 2012). All of these components need to be present in order for the habit to be successfully *practiced* over time. Drawing on the narratives in this chapter, a mom might: have the desire to teach her child about where food comes from (meanings); possess a spacious and sunny yard (resources); and know how to garden (skills). In this case, all of the components are in unison and she is afforded the ability to practice green parenting. In this example, the pro-environmental habit successfully emerged. Likewise, existing habits can deteriorate.

Practice and embodiment research on sustainability issues (Angelo 2013; Bhatti et al. 2009; Bonham 2006; Colby 2020; Macnaghten 2000a, 2000b; Marinho and Turini-Bruhns 2005; Roth and Bowen 2001; Whitman and Cooper 2000, 2011) is also important for showing how changes in policies and/or technologies can alter habits regardless of whether or not the practitioners hold pro-environmental beliefs. In this scenario, people are carriers and not initiators of change (Hards 2011). The data from the two bike studies illustrates this very clearly. While many cyclists do not commute for pro-environmental reasons, they are still reducing their ecological footprints by carrying out a green activity.

In addition, practice and embodiment research emphasize how the adoption of practices is facilitated by social modeling (Hendlin 2016). As Hal Wilhite (2015) explains, being exposed to a new practice can be even more powerful than simply being told about its benefits. When people see others doing something new, and then model it, they are learning with both their minds and bodies. Both the parenting and the choice commuter studies pointed to the importance of social modeling for enabling green habits (that is, fond childhood memories spent in the outdoors with families or seeing someone in the community successfully commute on a bike). Modeling an activity helps to normalize it and witnessing the activity transforms it from an idea into a plausibility. Trying it out for yourself, then, solidifies the idea into a potential embodied habit and you become the next role model. In a case study of people living sustainable lifestyles, Fiona Shirani and colleagues (2015, 72) found that practitioners served as *encounter agents* for fostering, in an indirect manner, sustainable change in the wider communities where such practices were not yet normalized. Like the practitioners in the Shirani et al. (2015) case study, green parents' narratives showed that they were aware of the importance of social modeling and used that knowledge to deliberately pass pro-environmental lifestyles on to others.

Finally, practice theory documents how major life events can enable new routines to arise (Darnton et al. 2011; Reckwitz 2002, 255; Sahakian and Wilhite 2014, 28; Shove et al. 2012; Tyers et al. 2019; Verplanken et al. 2011; Verplanken et al. 2008; Whitmarsh 2022). Recall that most parents experienced the welcoming of an infant into the home as an opportunity to ramp up their green practices. In addition, several of the pathways to green parenting (chapter 2) represent significant life events, including travel, higher education, taking a new job, and moving to a new neighborhood. Choice cyclists also named moving to a bikeable neighborhood as a pathway to commuting.

Practice theory, however, does not allow for a full exploration of motivation-driven human activity. For instance, while it does include meanings and skills in its understanding of how change takes place, it falls short in its ability to account for how people draw on green identities to strive (van Dam and Fischer 2013) to achieve agentic change (Hamilton et al. 2018) and skilled competence (De Young 2000). Recall, for instance, that 12% of choice commuters pointed to self-sufficiency (skilled mastery of the cycling machine and avoidance of dependence on a car) as a strong motivator for utility cycling. Mastery of the machine and avoidance of dependence on a car was a narrative that also showed up, in indirect ways, throughout many more of the choice and necessity commuter narratives. As I discuss below, innovation research complements social practice research by focusing attention on motivations to engage in deliberate behavior.

The innovation perspective (fully cited in the introduction) provides a useful framework for exploring how people sometimes deliberately innovate when conventional solutions no longer work. Innovation theory explains that, when faced with unanswered needs, people draw on meanings to craft novel solutions. Green parents, for instance, innovatively navigated around constraints (culture, infrastructure, money, time, and media) by drawing on holistic meanings of sustainable living to focus their practices on socializing their children into a better future. In other words, they were creatively imagining themselves and their children into a more sustainable world. Most imaginings die out before they become reality. As Josef Hochgerner (2011, 7) points out, "Social innovations are not solely determined by the potential of an idea, but also by whether and to what extent the potentials of an idea are realized."[10] Once a solution becomes realized, it represents a step toward dismantling the worldviews that caused the problems in the first place (Haxeltine et al. 2017). Innovative ideas appeared in parents' narratives when they articulated a discontent with habitual consumerism, loss of connection to nature and to each other, and an overall sense that something was missing in life. They drew on a set of meanings (consciousness, connectedness, and well-being) to address these problems and other challenges (culture, infrastructure, money, time, and media).

The efforts of parents are not only long-term but are also broad reaching. Their ideas are part of a bigger set of discourses in which people see pro-environmental beliefs as part of a larger holistic project (Breyer 2017; Dobernig and Stagl 2015; Gkargkavouzi et al. 2019; Hamilton et al. 2018; Hedlund-de Witt 2014; Kirby 2003; Lestar and Bohm 2020; Liftin 2016; Lorenzen 2012; Moisander and Pesonen 2002; O'Brien 2013; Prothero et al. 2010; Robertson and Lawrence 2015; Schlosberg and Coles 2016; Sutton 2020; Walther et al. 2016; Zylstra 2014). As Evans and Abrahamse (2009, 495) write, many collections of "people living green lives are promoting an alternative vision of . . . the good life." Discourses around the green good life are diverse but they do converge on promoting more holistic (Brown and Kasser 2005; Hedlund-de Witt 2014; Hedlund-de Witt et al. 2015; Helne and Hirvilammi 2015, 2019; Helne and Salonen 2016; Howell 2013; Gough 2017; Schlosberg and Coles 2016; Schor and Thompson 2014; Schor and Fitzmaurice 2015; Sterling 2011) approaches to well-being and critiquing consumerism (Binkley and Littler 2011; Rich et al. 2017; Schor and Fitzmaurice 2015; Soper 2007, 2011; Soper et al. 2009). Parents' focus on conscientiousness (Lorenzen 2012); connectedness (Dobernig 2015; Gkargkavouzi et al. 2018; Hedlund-de Witt 2012, 2014; Lorenzen 2012; Mayer and Frantz 2004; Stock and Forney 2014; Zelenski and Nisbet 2014; Zylstra et al. 2014); and well-being (Dobernig and Stagl 2015) and happiness (O'Brien 2013; Rich et al. 2017; Verdugo et al. 2011; Walther et al. 2016) have been documented among other people living green lives. Parents are participating in a broader movement to imagine a new worldview that includes, but is much more than, pro-environmental behavior.

Examinations of green lifestyles benefit from practice, innovation, and embodiment lenses. While the practice perspective provides insight on the ongoing mechanics of everyday change, the innovation perspective provides insight on intentionality (Schubert 2014). The embodiment perspective demonstrates how people experience and learn through doing.

The policy and programmatic implications of the findings of chapters 1–4 will be explored in the conclusion.

NOTES

1. Note that, in a 2020 article, where I focused on parents' long-term motivation to socialize their children into green lifestyles, I summarized only the top three motivations of footprint, socialization, and health as they were the most related to the theme of the paper (Auriffeille). Because, in this book, I am exploring the larger project of green living, I do also report on parents' motivations to exercise thrift. As I show in the sections on commuting, thrift is also an important motivator for cyclists.

2. I was surprised by parents' assertions that welcoming an infant into their households presented very few challenges to green living. Personally, I struggled to balance green living with raising an infant while pursuing a career. After rereading the interviews, I realized that the one parent that experienced green living as much harder, while raising an infant, was a working mom whose husband was stationed overseas. This resonated with my own history. For the first two years of Gaspard's life, my husband was working abroad. It is easy to do the math on this scenario; I was an exhausted, single working parent of an infant. I found myself doing things that were wasteful. As an example, Gaspard did not nap. He would nap in the car, but, only if the engine was running. I admit, I spent many a Saturday afternoon, parked in a shady spot, with the car engine running, while Gaspard slept. I likely spent that time grading my environmental sociology papers. When my husband transitioned to working/living in the United States, my dormant green habits gradually resprouted. Our household's habits evolved over time as we mapped out our sustainable lifestyles on a social, political, and physical terrain that is at times cooperative and, at other times, restrictive.

3. Studies have found that environmentally responsible behavior is positively impacted by participants' consideration of the future consequences of their behavior (Enzler 2013; Sabine and Bauer 2013).

4. In an earlier study, Fiona Shirani and colleagues (2013) explored the role of children/youth in enabling green parents to think themselves into a more sustainable the future. They found that the presence of children and youth in the household did help parents to make an ethical link to future generations and take a long-term view on climate change. However, they concluded that the difficulty for parents is not in

cultivating an ethical view, it is in maintaining that view in the face of daily parenting pressures (465). My study on green parenting (GP) also identified challenges for parents (see chapter 3). However, I argue that the challenges are temporary as parents focus on the longer-term goals of greening future generations.

5. One of the primary goals of Chawla's article was to integrate two areas of study: one, research on how young people cope with information about environmental devastation and, two, research on the benefits of connecting children to nature.

6. Here Louise Chawla (2014) is drawing on Nussbaum's (2011) capabilities approach to human development.

7. It is important to note that a number of studies are also exploring the positive impact that electronic media can have on promoting outdoor time and connections to nature. See for example, Joachim Schneider and Steffen Schaal's (2017) study on how smartphone game apps can foster connectedness to nature.

8. The literature delineates two categories of well-being. Eudemonic well-being is living a meaningful or purposeful life and hedonic well-being is associated with pleasure (Kahneman et al. 1999). I use the term in its broadest sense. Parents talked about meaningfulness and pleasure simultaneously.

9. Some cyclists made a point of clarifying that their current bike commuting is not strongly related to thrift (fourteen) or environmental concerns (ten).

10. These early imaginings often go unnoticed because most disappear before they manifest into change (Hochgerner 2011; Jaeger-Erben and Ruckart-John 2015, 163–64). Ideas that do lead to change go through a multiphase process. Early phases are hidden in people's everyday lives (Jaeger-Erben and Ruckart-John 2015, 163–64; Mulgan 2006). In later phases, ideas turn into action and experimentation takes place. In the third and fourth phases, after experimentation, the innovation grows and, if successful, becomes normative and then evolves (Mulgan 2007).

Conclusion

The main thread that runs throughout this book is that people's everyday lives matter for long-term local and global sustainability. Large-scale environmental policies, after all, evolve out of people's everyday doings and exchanges of symbols.[1] People come to sustainable lifestyles from multiple, overlapping experiences. Sometimes, as with parents, practitioners were driven by green ideals to scale up their pro-environmental habits. For others, like many bike commuters, pro-environmental beliefs evolved out of the everyday doings of cycling. As I discussed in chapters 2 through 4, the practice, innovation, embodiment, and attention economy frameworks help illuminate the ways that pro-environmental behaviors are experienced and shared, and how they emerge, flow, maintain, evolve, and die out. The terrain of sustainable living is, as Lisa Siegel and colleagues (2018) conceptualized, a *living forest*.

Understanding green lifestyles as a *living forest* demands that we abandon simplistic formulas, like providing consumers with endless data (toxicity information on a product label, for example). We must engage practitioners at all levels: the meanings they hold, the resources they have access to, the skills and knowledge they have developed, and their emotional and embodied experiences. Governmental and social change efforts must shift from a moralistic focus on encouraging better consumer choices to a more interdisciplinary focus on enabling everyday socially shared practices for people with and without pro-environmental beliefs. It also implies that regional, national, and global systemic-level policies must be informed by people's daily life experiences. Interventions and policies should observe what people are already doing, examine their pre-existing strengths, and ask what brings them joy. The pleasures of spending outside time with children or of biking through a rainstorm on the way home from work, for instance, should be looked at as critical entry points for pro-environmental intervention and policy making.

Understanding green lifestyles as a *living forest* suggests that advocates seek out interdisciplinary research from the social, applied, and natural sciences to craft programs and policies that enable pro-environmental practices.

As such, this book has drawn on research and theory from a mix of disciplines including sociology, psychology, philosophy, and urban planning, among others. Below, I pull from this diverse and overlapping body of work to suggest emergent policy and programmatic directions.[2]

POLICY IMPLICATIONS

How can we bring about a more environmentally and socially sustainable future? This book, along with other works on green living cited throughout the preceding chapters, are only beginning to scratch the surface of how to enable people to successfully live sustainable lives. The narratives of parents and cyclists point to several key areas where we should focus our research, and policy, programmatic, and infrastructural interventions.

Practitioners named a number of piecemeal and overlapping pathways to green living (such as travel and exposure to other cultures for parents [GP], the motivation to exercise for choice cyclists [CB], and economic need for necessity cyclists [NB]) and pointed to specific challenges they faced in their endeavors, for example, poor alternative transportation infrastructure for respondents in all three studies. Parents and cyclists also discussed a number of motivations (such as convenience for cyclists) and meanings (such as wanting to make conscientious decisions for parents) that fueled their green behaviors. Out of this macrame of pathways, challenges, motivations, and meanings, practitioners engaged in successful pro-environmental change. Their experiences point to several mutually reinforcing enablers of change that sustainability social science researchers, advocates, and policy makers should learn more about and do more to incorporate into interventions.

First, we need to expand efforts to *educate the whole person*, regarding both children and adults (Clover 2002; Scopelliti et al. 2022) and not just the consumer. Parents pointed to a range of complex motivations and meanings, and cyclists, likewise, voiced a variety of motivations for engaging in pro-environmental behavior. Green parents, for instance, desired family well-being (GP) and cyclists desired the simple bliss of commuting to the store on a bike (CB and NB). They found green activities to be rewarding. But, sometimes, infrastructure, money, time, skills, or other challenges got in the way. In addition, as I showed with the green parenting study, many practitioners wanted to learn about environmental issues and practices and were eager to partake in deliberate decision making and engage in critical analysis of environmental media. All of this points to the need to create education programs that foster space for critical engagement, knowledge creation, and skill development, and that enable people to connect to each other and nature in meaningful ways and work together to foster personal and ecological well-being.

Two examples of adult environmental education initiatives that focus on the whole person, that I did not discuss in the green parenting analyses but that did appear in the interview data, are locally run outdoor programs. Both organizations have been founded in the last 10 years and have national branches but are administered by the residents that live in the areas around their regional offices. One parent, April (1 child, I-41, GP) brought up the importance of her family's participation in the regional *Outdoor Afro* group.[3] She talked about how Outdoor Afro organizes hikes and outdoor excursions for anyone to attend. Occasionally the excursions involve learning about local environmental or social history. Each regional group focuses on its own unique resources, culture, and history. Several green parents (GP) also brought up the Charleston area *Hike it Baby* group.[4] Parents connect through the website and meet up for baby-, toddler-, and child-friendly excursions. Often there is someone in the group who is able to identify plant life, for instance. These types of groups provide ways for people to get outside, talk to each other, informally educate and model green behavior, and so on.

Finally, this was not brought up in the interviews, but when I was recruiting for the green parenting (GP) study, I attended a workshop on how to make green cleaners out of citrus peels and vinegar. The workshop was hosted by the downtown Charleston hostel, *Not So Hostel.*[5,6] The manager of the hostel is a local environmental advocate and practices pro-environmental behavior in the daily operations of running the hostel. She also regularly hosts small workshops on green living. I found the workshop to be very helpful for skill building. I'm still making household cleaners out of citrus peels and vinegar and was surprised to find how simple it is. As many of the green parents and cyclists pointed out (GP, CB, and NB), some of the green activities are really not that hard; you just need investment time, like attending a workshop on making green cleaners, to learn them. I think this workshop style was also successful because, as opposed to social media, I wasn't just learning about making cleaners, it was modeled for me. Also, the workshop took place in a living room setting and a local green-focused restaurant owner attended and brought sandwiches. This made it fun and helped to indirectly reinforce that green living does not have to be about sacrifice, it can be about connecting to others and enjoying a bit of conversation about sustainability issues. These are just a couple of local examples that I consider to be successful at educating the whole person. They are not meant as models but are posed as suggestions for how to move forward with what, no doubt, must be countless varieties of opportunities to offer green education experiences for adults.

Second, *identify household-level shared values*. The green parenting study pointed to enormous potential for policies to tap into shared values among the diverse members of a household. Recall that parents (GP) and choice bikers (CB) talked about early family experiences and role models as important influences on their adult pro-environmental behaviors and/or values. Parents shared key environmental goals (socialization of children, among other things) and meanings (well-being, for example) and this helped them work together to overcome short-term challenges. Socializing children into green living and connecting to nature appears to have positive short- and long-term benefits on lifelong pro-environmental behavior and beliefs (fully cited in chapter 4). However, households also, inevitably, represent terrains of competing interests and individual identities. The families in the green parenting study either still had young children or recently had young children in their households. As their children enter adolescence, conflicts over energy use and other green practices might arise (Collins 2015). I see this in my own family as our son, Gaspard, grows up. When he was a toddler, I delighted in hearing him point to mega-sized polluting cruise ships in the Harbor and declare "poo poo ships!" to anyone who would listen. Most of his clothes were hand-me-downs and he never noticed or cared. He is still a thoughtful environmentalist who actively models our green practices and eagerly brings home information from school about sustainability. However, now that he is a ten-year-old basketball fanatic who styles his hair first thing in the morning, even before going to the swimming pool, and keeps his court shoes in their original box, he has repositioned himself as a self-aware, brand-conscious consumer. Practices, like buying second-hand clothes, that I used to take for granted are now negotiated. In sum, programs need to avoid overly individualistic and consumeristic approaches to green change and focus more attention on households and the evolving relationships, interests, differences, and shared values within those households (Henwood et al. 2016; Jamieson 2016; Webb et al. 2016). As Lynn Jamieson writes, in her piece on families, relationships and sustainability, "families and personal relationships are integral to 'world making' . . . " (2016, 346).

Third, *get outside*. Green parents and cyclists (GP, CB, and NB) voiced what a growing body of evidence shows, getting outdoors and connected to nature and the practice of pro-environmental behaviors enhance both overall well-being and feelings of happiness (fully cited in chapter 4). Current green campaigns and some social science research focus too much on the story of sacrifice and deprivation (Brown and Kasser 2005). This study shows that, rather than a sense of sacrifice, green practitioners are driven by the desire for well-being, connection, and happiness. Advocates should tap into people's corporeal pleasures (desire to exercise, socially bond, etc.) to get them started on green routines.[7] This is not to say that practitioners are unwilling to make sacrifices. Several studies connect willingness to sacrifice to engagement in green behaviors (Han et al. 2018; Knez 2016; Mostafa 2013; Tsai 2021).

Rather, the greater point is that, while willingness to sacrifice may be a common trait held by people who engage in green behavior, important drivers (Shyang-Chyuan 2020/2021), and also outcomes (Corral-Verdugo 2011; Rich et al. 2017; Prinzing 2020), of green living are pleasure, happiness, well-being, etc. For practitioners who do not intend to engage in pro-environmental behavior, willingness to sacrifice is a non-issue. For green-minded practitioners, they may hold a willingness to sacrifice but this does not mean they experience their green lifestyles *as a sacrifice*. In sum, environmental advocates and policy makers should explore ways to highlight the positive outcomes of green living. The three initiatives that I mentioned above (*Outdoor Afro*, *Hike it Baby*, and *Not So Hostel*) all provide good examples for engaging practitioners in rewarding activities where they can connect to each other and/or nature while sharing new green habits.

Fourth, *make it accessible and make it normal*. As social practice research documents, practitioners often engage in pro-environmental behaviors without intending to be green (fully cited in the introduction). In addition, both intended and unintended green practices can potentially lead to positive spillover (fully cited in chapter 4). For instance, as I documented with the choice cycling interviews (CB), commuting for exercise frequently led practitioners to adopt other pro-environmental behaviors/attitudes. However, getting people on their bikes to commute can be challenging in the Lowcountry as participants in all three studies (GP, CB, and NB) pointed to poor bike/pedestrian infrastructure as a significant barrier to engaging in pro-environmental behavior. Participants in all three studies also pointed to the need to normalize pro-environmental behavior, like bike commuting. In particular, the choice and necessity cyclists (CB and NB) described how both an anti-bike culture and poor physical biking infrastructure served as commuting challenges. As I discussed earlier in the book, I live only a couple of miles from the College of Charleston campus and it is an absolute shame that I am unable to bike to work. If it were safe, I would certainly do it.

How can cities get people, who would like to bike commute to work but find it inaccessible, on their bicycles on the roads? Mounting evidence out of diverse interdisciplinary fields points to the immense power of urban design to make cycling more accessible, including bike commuting (for just a few examples see: Claus 2012; Dehghanmongabadi and Sebnem 2020; Handy et al. 2002; Hull and O'Holleran 2014; McMillan 2007; Nielsen et al. 2013; Panter et al. 2008; Polanski 2019; Raccioppi et al. 2008; Titze et al. 2008; Zhao 2014).[8] Municipalities need to take advantage of this evidence and start prioritizing the legitimate rights of cyclists to be on the roads and designing safer streets and bridges for bikers and pedestrians. In addition, as the bike commuters in both cycling studies pointed out (CB and NB), it is not simply about installing infrastructure; cities and towns need to enforce a local transportation culture that is supportive of cycling and walking. Models for how to do this are beginning to emerge (Craig 2008; Hochli et al. 2019; Lugo 2013; Nikitas 2018; Rose and Marfurt 2007; Patricio and Kruszielski 2016; Rose et al. 2016). Some research shows that simply increasing the number of cyclists on the streets leads to safer roads for everyone and results in fewer accidents (Elvik and Goel 2019; Marshall and Garrick 2011; Marshall and Ferenchak 2019; Nordback et al. 2014). Possible programs to get more people cycling include holding bike-to-work days (Hochli et al. 2019; Marfurt 2007; Patricio and Kruszielski 2016) and starting Bike Share programs (Nikitas 2018). These types of initiatives, by getting more bikes on the roads, normalize biking and foster bike-friendly cultures.

Fifth, *avoid guilting people*. Programs need to avoid overly moralistic messages about "being good environmentalists." This is especially true when infrastructure (say bikeable roads) for making green choices is substandard. In addition, research suggests that guilt, and other negative emotions, often fail to motivate people to engage in pro-environmental behavior (Bissing-Olson et al. 2016; Rubens et al. 2015). This study on Lowcountry green lifestyles demonstrates that, rather than guilt, many green practitioners are driven by the desire for well-being, connection, pleasure, and happiness and also by practical considerations like bike commuting in order to save money and avoid traffic/parking hassles. Advocates should tap into people's corporeal pleasures (desire to exercise, socially bond, etc.) to weave a compelling story about the delights and joys of bike commuting, going on a hike with your child, starting a food garden, etc.

Sixth, *promote exposure to other cultures*. Advocates need to build off of practitioners' reports that they are responding positively to exposure to other green cultural habits through travel or moving homes. Environmental organizations, for instance, should look into funding educational components of study abroad programs, alternative breaks, etc. as a way to grow sustainability awareness. While travel itself, especially air travel, is a significant contributor to air pollution and climate change, advocates might think about piggybacking environmental education on programs that are already in place, like college study abroad courses, rather than starting new programs. In addition, people in the twenty-first century are extremely mobile. We sometimes are required to move homes for career, family, etc. Cities and towns should think about ways they might reach out to new residents to tap their unique knowledge and habits.

Seventh, pay attention to *moments of change*. This study supports research which documents that key life changes (say, becoming a parent, attending a college, entering retirement, moving to a new neighborhood, etc.) can serve as moments of change when new habits can be started or enhanced (Reckwitz 2002, 255; Sahakian and Wilhite 2014, 28; Shove et al. 2012; Verplanken et al. 2011; Verplanken et al. 2008). Such moments of change, like becoming a new parent, serve as extremely fertile grounds for tapping into green beliefs and behaviors that people are already practicing. Programs should explore ways to leverage these key moments.

Eighth, *identify unique challenges and interests at the personal, household, and community level.* Lowcountry parents (GP), named time, money, media, conservative culture, and substandard green infrastructure as barriers to scaling up their green living practices. Choice and necessity bikers (CB and NB) named anti-biking culture and substandard biking and public transit infrastructure as challenges. Some choice cyclists (CB) and all necessity cyclists (NB) pointed to economic necessity as initiators of their commuting habits. Many of the parents and choice and necessity cyclists (CB and NB) pointed to thrift, among many other desires, as a current interest. Programs should put more energy into identifying specific practitioners' challenges and interests so that they can target programs to their unique life requirements. For the sake of brevity, I'll focus on the interviewees' stated interests in thrift. This study on Lowcountry sustainable lifestyles echoes other research that has reported thrift as a commonly held interest by a variety of green practitioners (among both those who hold pro-environmental identities and those who don't) (Corral-Verdugo et al. 2015; Gao 2018; Hitchings et al. 2015). Given the common reference to money and thrift in this particular study's narratives, appealing to potential practitioners from the angle of cost-savings, for instance, might give an environmental advocacy organization additional leverage to encourage pro-environmental behaviors. For instance, the *Not So Hostel* workshop I attended on making green cleaners (described above) was solely advertised as a sustainable lifestyle event. It could have easily played up the cost-savings component as well and might have, in doing so, drawn a more diverse set of attendees.

Ninth, *simplify media information and scale up social modeling.* As I discussed in earlier chapters, parents talked about having a complex relationship with media (GP). Some parents were successful at using it to enhance their own pro-environmental habits and knowledge. Others found it to be confusing and misleading. These findings are in line with research which shows that modeling green behavior is more powerful than providing people with information (Sussman and Gifford 2013). Seeing someone successfully cycle to the store, with a child in a bike-seat, is a more impactful producer of change than social media tweets (the strength of social modeling for producing pro-environmental change is fully documented in chapter 4). Green campaigns need to create more opportunities for modeling and rely less on media-based messaging. The *Not So Hostel* nonprofit discussed above, and the *Fresh Future Farm* nonprofit discussed below, provide good examples of organizations that focus on modeling pro-environmental behavior. This is not to say that there is no role for media. I found out about the *Not So Hostel* soap-making workshop through social media. Rather, advocates need to try to shut down some of the media "noise" and work to produce more practical information that is understandable and useful.

Tenth, advocates need to *get green environmental politics out* of campaigns for sustainable living. Analyses of the interviews with parents (GP) and cyclists (CB, NB) point to how people come to pro-environmental behaviors from all kinds of political orientations and cultures. Recall that many choice cyclists (CB) started commuting for the exercise and only came to hold pro-environmental ideas later, as a result of bike commuting. Organizations may hold very specific political agendas along with environmental goals, but the research in this book on Lowcountry residents living sustainable lifestyles suggests that green advocates should tone down their political messaging and focus more on disseminating and modeling practical pro-environmental habits. This suggestion is in line with the findings of other studies which document green lifestyles among people that do not profess strong or any pro-environmental politics (Colby 2020; Donald et al. 2010; Jarosz 2000; Maye et al. 2007; White 2011).

Finally, *embrace agency*. As the social innovation research highlights (fully cited in the introduction), when societal policies and institutions fail to enable necessary change humans deliberately innovate, in their daily lives, new ways of doing things. Some cyclists' (CB and NB) and all parents' (GP) narratives told stories about pro-environmental intentionality. Practitioners who hold green beliefs want to act on them and feel the sense of competence that comes from making a positive difference. This need on the part of sustainably minded practitioners to enact agentic change has been documented in other research as well (De Young 2000; Hamilton et al. 2018; Lorenzen 2012). Advocates should develop programs that encourage practitioners to engage in problem solving and solution generation. As positive examples for how advocates might encourage deliberate and innovative action, below I discuss two local nonprofit organizations that each successfully engaged in agentic change.

Collective action for bike/pedestrian and public transit has been quite strong in Charleston for several decades. Ron (I-35, CB), one of the choice cyclist interviewees, and a local advocate for bike/pedestrian and public transit issues, recounted the story of how the Ravenel Bridge got its much-beloved and widely utilized Wonders Way bike-pedestrian path. Leading up to the construction of the Ravenel Bridge before it was opened in 2001, the city of Charleston and the town of Mt. Pleasant, which occupy the two sides of the bridge, were negotiating the shared costs for the inclusion of the path. They were having trouble coming to an agreement and it began to look like the path might be excluded from the proposed bridge plans. As Ron told the story, a local bike/pedestrian and active transit organization (what is now *Charleston Moves*[9]) had been keeping an eye on the developments of the proposal for a bridge path and responded to the shaky negotiations in an innovative way. As Ron recounted, the organization took out a full-page advertisement in a local paper in which they thanked the town and city for agreeing to include the bike/pedestrian path in their bridge plans. Area citizens were thrilled to read that the two local governments had decided in favor of the path and they flooded municipal offices with excited *thank you* phone calls. In the end, likely as a result, at least in part, of the advertisement, the path was built and now commuters, exercisers, and tourists enjoy and utilize Wonders Way path every day, all day.

Below I discuss another local nonprofit, *Fresh Future Farm*, that emerged out of the deliberate vision of its founder, Germaine Jenkins, to, first, provide her family with sustainably sourced food and, later, to share her knowledge and skills with the greater community. The organization is a Lowcountry environmental initiative that not only illustrates the important impact that individual and communal agency can have on sustainability, but it also draws on many of the above suggestions, like educating the whole person, getting people outside and connected to nature and each other, engaging in social modeling, etc. In addition, Germaine utilizes the farm to connect local practitioners, policy makers, and advocates over systemic environmental justice and food access issues.

FRESH FUTURE FARM

Fresh Future Farm[10] is a nonprofit urban farm, grocery store, and community outreach organization. Germaine, one of the interviewees of the green parenting study (3 children, I-48, GP), earned her culinary degree as a single parent while supporting her children on public assistance and later earned her Master Gardener Certificate through a Clemson Extension course. After earning her degree and certificate, she started the solar and wood-chip-powered organic urban farm and grocery store in a low-income food desert in a North Charleston community. The farm produces local organic produce which the grocery store sells, on-site, to people in the neighborhood and visitors. Parents and children from all over the greater Charleston region come to the farm to volunteer and also learn about organic and sustainable food production. The farm serves as an outdoor community center where people can come together to volunteer, role model, learn, and connect to each other and nature.

The farm also serves as an important demonstration of *hope in action* to members of the low-income, largely African American community. Germaine talked about how:

> I've seen changes . . . at the community garden. Once we started to beautify on what was once the trailer park before the city bought it . . . and make compost and plants available, people started beautifying their own yards. We [donate compost] in the summer and it's gone in a month. People will just empty their trash bins, fill up, and then roll it on home, 'cause you make do [laughs]. Some of them gentlemen that did that got awarded best yard a couple of months in a row, and then, it became a competition. People near [them] also wanted to improve. I think beautification is contagious, and we just wanna share the no-chemical piece of it. . . . I think communities with high levels of poverty, there's less hope, but you see cars drivin' down the street [next to the farm] that almost screech to a stop because all of a sudden, there's banana trees and blackberries and blueberries, things you don't typically see. Just that is enough to change somebody's attitude, outlook, all those things. (I-48, GP)

She continued to talk about how important the farm can be for modeling to others that sustainable living is possible. She noted:

> I wanna put solar panels on this buildin' . . . so that we can do classes and demonstrate a lotta different techniques. 'Cause once I show some guys that live in the neighborhood [that] I can make hot water with [recycled wood chips], that's gonna change some people's thinkin' . . . for people to really see that you can cook beans [for five families] on this oven that we created in the middle of the community garden [at the time of the interview, Germaine hadn't yet installed some of the energy-efficient technologies, like solar panels, that are now in place]. (I-48, GP)

Germaine's story about the importance of beautification echoes one of the main threads woven throughout this book. Green living is about more than reducing ecological footprints. It is about living a good life, or a *beautiful life*.

Germaine, and her family and staff, also visit area nonprofits, campuses, and schools to talk about environmental justice issues, food security, and sustainable living. Germaine spoke about the importance of sustainability education in the community and schools and explained:

> I need to get to the people that don't see it the way I do. . . . [We need to] reach out to neighboring schools so that kids can go and start yappin' at their parents, "I learned this today. You can save this, da da da da da, trash is bad," especially in a neighborhood like ours where people are sometimes throwin' trash in the street and it impacts water. I wanna make these connections with the kids so they can pass it on to the parents. (I-48, GP)

This conclusion presented a brief discussion of a few successful nonprofit initiatives, like *Fresh Future Farm,* that point to paths for encouraging more people to successfully engage in green habits. I also listed numerous ideas, generated out of the Lowcountry green living interview data, to suggest ways to improve sustainability policy and advocacy. However, I ultimately presented more questions than answers. While the analyses suggest ways to improve efforts to encourage sustainable habits, much remains to be known about how to actually realize those ideas into programmatic practice.

WHERE DO WE GO FROM HERE?

Future applied and interdisciplinary social science and urban planning research on sustainability should ask many questions raised by these results. How can advocates craft programs that *educate the whole person* and how do people respond to education programs that draw on their holistic understandings of environmentalism? How can education programs and other organizations encourage more conscientious decision making around green issues? How can advocates connect people, like Lowcountry practitioners, with other communities of people living green lifestyles and who hold similar worldviews that focus on holistic meanings like well-being? Is there potential for this holistic worldview to energize collective action for social and environmental sustainability? If yes, how do we foster this activism in people's daily lives and, also, in the public sphere? Which travel and cultural experiences produce the most improvements in pro-environmental behavior/beliefs? How can advocates reconcile the positive impact that travel has on pro-environmental behaviors/beliefs with the enormous negative impact that air travel, especially, has on climate change (UNEP 2021b)? How can organizations devise campaigns that appeal to people across the entire political spectrum? What kinds of outdoor embodied experiences best enable pro-environmental changes in beliefs and behaviors? How can cities and towns develop their natural resources to get people outside and how can they make those spaces accessible to everyone? What manner of childhood pro-environmental exposure and education is most successful in producing lifelong green habits? In addition, while all big life changes seem to offer moments where habits can evolve, in terms of sustainability, having a baby appears to be an extremely important moment. How can advocates leverage this important moment of change to help parents continue and increase their sustainable habits? How can programs help practitioners overcome the household-level challenges of time, money, and media navigation? How can advocates help practitioners with finding *investment time* to start a new routine? How can programs help household members work together to negotiate their habits in pro-environmental directions? How can advocates produce green media that is accessible, clear, simple, and still accurate and how can they help practitioners quickly filter through inaccurate information to identify accurate information? How can programs best incorporate social modeling into their interventions? What infrastructural improvements work best in specific communities? Peninsula Charleston has very narrow streets, for instance. Bike paths are impossible on most downtown roads. How can infrastructure in this type of historic city be improved for cyclists? When municipalities lack green infrastructure, like transit, how best should they incorporate citizens' habits, needs, and interests into future plans? In cases where the challenge is cultural, like the anti-bike culture in Charleston, how can governments create a supportive bike commuting culture?

Finally, how can organizations enable green-minded practitioners to engage in more agentic practices and develop a sense of competence? For instance, Germaine Jenkins' story of how she started and grew *Fresh Future Farms* is extraordinary. That kind of agentic environmental innovation should, instead, be ordinary. How can advocates help other practitioners launch their ideas into realities?

These are a just a few of the questions we need to ask as we move forward.

ONE LAST THOUGHT

Thinking about this question of how to enable green lifestyles, I often reflect back on a conversation I had with Zoe, one of my student researchers. Zoe and I were combing through the green parenting data and the emerging results inspired her to share a story that her family tells about her paternal grandparents. They lived in Brooklyn and Zoe's grandmother would wash the family laundry in a bucket, in the back of the yard, and hang it on a line. As a surprise, Zoe's grandfather splurged and bought her grandmother an electric washing machine. They all joked about how she never used it because she would have missed her social time with all her friends, hanging out at their back fences, while doing the laundry by hand.

For me, this story perfectly captures many of the narratives woven throughout this book. Habits are alive; they are practiced in people's daily lives and imaginations. They evolve and grow and, sometimes, decay. Zoe's grandmother, like all of the respondents in this book, enacted her daily life within a mix of personal and social identities that crisscrossed resources (such as the money to buy an electric washing machine) and skills (such as knowing how to clean laundry, by hand, in a bucket). Our policies and programs need to do better to meet people where they really are. What matters to them? What are they already good at doing?

For too long we have allowed ourselves to buy into the false idea that we can wave the provision of pro-environmental information, like a magic wand, to fix our planet's sustainability problems.[11] Zoe's grandmother engaged in what we now call *pro-environmental behavior*, washing the family's clothes by hand and hanging it all on a line to dry. While she wasn't intentionally practicing a green behavior, she did know what many of the participants in these three Lowcountry interview studies also expressed, that social connection and feelings of pleasure and well-being are fundamental to healthy human existence. This simple fact is where our future research agendas and policies need to begin. This simple fact explains why parents are so critical of green media and consumeristic approaches to green living, and it also explains why bike commuters would prefer to joyfully ride through a torrential rain downpour than sit trapped in a car listening to, often moralistic, news about how we all need to make more *sacrifices* in order to enact sustainability. Our policies need to begin with people's embodied embeddedness in their social, imaginational, and physical homes. Tom Hargreaves (2011, 94) conceptualized this kind of sustainability as bringing change from the "inside-out."

NOTES

1. By focusing on everyday sustainable practices, I am answering the call of several sociologists that it is time for environmental sociology to branch out of its longtime focus on macro-level political economic issues to look, also, at micro-level social interaction (Brewster and Bell 2009; Hargreaves 2016).

2. While the scope of this book does not draw on the natural sciences nor on applied technological research (except in references to the need to engineer better green infrastructure like bike/pedestrian friendly roads and paths), it is implied that advocates must also draw on these disciplines to enact green change.

3. https://www.meetup.com/Outdoor-Afro-Charleston-SC/?_cookie-check=7f-86P472JWtnN9q

4. https://community.hikeitbaby.com/branches/charlestonlowcountry-sc/

5. https://notsohostel.com/press

6. While I did not recruit any parents at this workshop, it did provide first-hand experience of what interviewees were telling me about the importance of investment time to learn a new habit that may end up being simple to perform in daily life.

7. Miranda Mirosa and colleagues (2011) found that pleasure was associated with some pro-behaviors, like hanging laundry out on the line. However, they also found that pleasure was associated with other resource-inefficient behaviors, like taking a long shower. I agree that pleasure-seeking can lead to wasteful practices. However, my focus is more on how advocates should focus campaigns and programs on helping

people to see that green living can be enjoyable and also on helping people to realize pleasure in their pro-environmental lifestyles.

8. For numerous reports on how urban planning can positively impact utility biking and walking, and also public transit use, see publications in the *International Journal of Sustainable Transportation.*

9. https://charlestonmoves.org/

10. https://www.freshfuturefarm.org/

11. In this book I have juxtaposed media with modeling. However, while pro-environmental technologies are only indirectly broached in this book (say, when I refer to sustainable transportation infrastructure), it is important to note that purely technology-driven, like purely information-driven, approaches to green living suffer from the same problems. They represent disciplinary silos when we need multi-disciplinary webs. We are increasingly learning that such methods, on their own, are often not successful and, sometimes, lead to solutions that are worse than the original problem. The disastrous history of recycling as a *solution*, especially the recycling of plastics, is a case in point (Mahdawi 2020). We cannot just hand humans a technology, like plastics recycling, or information, like the data on sea life killed every year by single-use plastics, and expect practitioners to successfully adopt new green habits.

References

ABC News. 2021. “Town of Summerville Expands to Single-Stream Recycling.” *ABC News*, April 29, 2021. Accessed March 22, 2022. https://abcnews4.com/news/local/town-of-summerville-expands-to-single-stream-recycling.

Adonia, E. Lugo. 2013. “CicLAvia and Human Infrastructure in Los Angeles: Ethnographic Experiments in Equitable Bike Planning.” *Journal of Transport Geography* 30: 202–7.

Afflerback, Sara, Shannon K. Carter, Amanda Koontz Anthony, and Liz Grauerholz. 2013. “Infant-Feeding Consumerism in the Age of Intensive Mothering and Risk Society.” *Journal of Consumer Culture* 13 (3): 387–405. https://doi.org/10.1177%2F1469540513485271.

Alexander, Samuel, and Simon Ussher. 2012. “The Voluntary Simplicity Movement: A Multi-National Survey Analysis in Theoretical Context.” *Journal of Consumer Culture* 12 (1): 66–86. https://doi.org/10.1177/1469540512444019.

Anderson, Monica. 2017. “For Earth Day, Here’s How Americans View Environmental Issues.” *Pew Research Center: FactTank News in Numbers*. April 20, 2017. Accessed January 29, 2019. https://policycommons.net/artifacts/617925/for-earth-day-heres-how-americans-view-environmental-issues/1598805/.

Angelo, Hillary. 2013. “Bird in Hand: How Experience Makes Nature.” *Theoretical Sociology* 42: 351–68.

Armstrong, James B., and James C. Impara. 1991. “The Impact of an Environmental Education Program on Knowledge and Attitude.” *The Journal of Environmental Education* 22 (4): 36–40.

Arp, Wiwam, and Christopher Kenny. 1996. “Black Environmentalism in the Local Community Context.” *Environment and Behavior* 28: 267–82.

Atkinson, Lucy. 2012. “Buying In to Social Change: How Private Consumption Choices Engender Concern for the Collective.” *Annals of the American Academy of Political and Social Science* 644 (1): 191–206. https://doi.org/10.1177/0002716212448366.

Auriffeille, Deborah McCarthy. 2020. “Before She Was Born, I Ate Cheerios and Beer for Dinner: A Qualitative Examination of Green Parenting in Lowcountry South Carolina.” *Humanity & Society* 45 (4): 439–70.

Austin, Kemen G., Mariano Gonzalez-Roglich, Danica Schaffer-Smith, Amanda M. Schwantes, and Jennifer J. Swenson. 2017. "Trends in Size of Tropical Deforestation Events Signal Increasing Dominance of Industrial-Scale Drivers." *Environmental Research Letters* 12 (5). Accessed March 19, 2022. https://iopscience.iop.org/article/10.1088/1748-9326/aa6a88/meta.

Averett, Kate Henley. 2021. "Queer Parents, Gendered Embodiment and the De-essentialization of Motherhood." *Feminist Theory* 22 (2): 284–304.

Backhaus, Julia, Audley Genus, Sylvia Lorek, Edina Vadovics, and Julia M. Wittmayer, eds. 2017a. *Social Innovation and Sustainable Consumption: Research and Action for Societal Transformation*. New York: Routledge.

Backhaus, Julia, Audley Genus, and Julia M. Wittmayer. 2017b. "Introduction: The Nexus of Social Innovation, Sustainable Consumption and Societal Transformation." In *Social Innovation and Sustainable Consumption: Research and Action for Societal Transformation*, edited by Julia Backhaus, Audley Genus, Sylvia Lorek, Edina Vadovics, and Julia M. Wittmayer, 1–11. New York: Routledge.

Bakir-Demir, Tugce, Sibel Kazak Berument, and Basak Sahin-Acar. 2019. "The Relationship Between Greenery and Self-Regulation of Children: The Mediation Role of Connectedness." *Journal of Environmental Psychology* 65. Article 101327.

Ball, Kylie, Adrian Bauman, Eva Leslie, and Neville Owen. 2001. "Perceived Environmental Aesthetics and Convenience and Company Are Associated with Walking for Exercise among Australian Adults." *Preventive Medicine* 33 (5): 434–40. https://doi.org/10.1006/pmed.2001.0912.

Ballantine, Roy, and Jan Packer. 2011. "Using Tourism Free-Choice Learning Experiences to Promote Environmentally Sustainable Behavior: The Role of Post-Visit 'Action Resources.'" *Environmental Education Research* 17 (2): 201–15. https://doi.org/10.1080/13504622.2010.530645.

Barr, Stewart. 2006. "Environmental Action in the Home: Investigating the 'Value-Action' Gap." *Geography* 91 (1): 43–54.

Barrera-Hernandez, Laura Fernanda, Mirsha Alicia Sotelo-Castillo, Sonia Beatriz Echverria-Castro, Cesar Octavio Tapia-Fonllem. 2020. "Connectedness to Nature: Its Impact on Sustainable Behaviors and Happiness in Children." *Frontiers in Psychology* 11. Article 276.

Barron, Jennifer. 2016. "Community Gardening: Cultivating Subjectivities, Space, and Justice." *Local Environment: The International Journal of Justice and Sustainability* 22 (9): 1142–58. https://doi.org/10.1080/13549839.2016.1169518.

Bartelme, Tony. 2020. "A Sunny Day in Charleston and a Flood: What that Tells Us About Climate Change and the Future." *Post and Courier*, September 22, 2020, A1. Accessed March 22, 2022. https://www.postandcourier.com/rising-waters/a-sunny-day-in-charleston-and-a-flood-what-that-tells-us-about-climate-change/article_367fb068-f9b9-11ea-b881-4ff1fcafa0a9.html.

Barton, Jo, Rachel Bragg, and Jules Petty. 2016. "The Wilderness Expedition: An Effective Life Course Intervention to Improve Young People's Well-Being and Connectedness to Nature." *Journal of Experiential Education* 39 (1): 59–72.

Bashford, Alison. 1998. *Purity and Pollution: Gender, Embodiment, and Victorian Medicine.* London: Palgrave.

BCDCOG (Berkeley-Charleston-Dorchester Council of Governments). 2005. *Regional Pedestrian and Bicycle Action Plan*. BCDCOG, Charleston, SC.

Behre, Robert, and Abigail Darlington. 2018. "Charleston Area Deadly for Cyclists and Pedestrians, but Change Is at the Mercy of SCDOT." *Post and Courier*. November 1, 2018. Last updated January 25, 2021. Accessed June 21, 2022. https://www.postandcourier.com/news/charleston-area-deadly-for-cyclists-and-pedestrians-but-change-is-at-the-mercy-of-scdot/article_ac06e51a-dc4f-11e8-9125-a7568b06f95b.html

Benesh, Sean. 2014. *The Bohemian Guide to Urban Cycling*. Skyforest, CA: Urban Loft Publishers.

Bennett, Tony, Francis Dodsworth, Greg Noble, Mary Poovey, and Megan Watkins, eds. 2013. "Special Issue: Habit." *Body & Society* 19 (2–3).

Benzecry, Claudio E. 2009. "Becoming a Fan: On the Seductions of Opera." *Qualitative Sociology* 32 (Jun): 131–51. https://doi.org/10.1007/s11133-009-9123-7.

Berger, Ronald J. 2008. "Agency, Structure, and the Transition to Disability: A Case Study with Implications for Life History Research." *Sociological Quarterly* 49 (2): 309–33. https://www.jstor.org/stable/40220070.

Bergman, Noam, Nils Markusson, Peter Connor, Lucie Middlemiss, and Miriam Ricci. 2010. "Bottom-Up, Social Innovation for Addressing Climate Change." *Sussex Energy Group Conference (ECEEE)*. Brighton, UK. Accessed February 1, 2019. https://www.researchgate.net/publication/271214929_Bottom-up_social_innovation_for_addressing_climate_change.

Bhatti, Mark, Andrew Church, Amanda Claremont, and Paul Stenner. 2009. "'I Love Being in the Garden': Enchanting Encounters in Everyday Life." *Social and Cultural Geography* 10 (1): 61–76. https://doi.org/10.1080/14649360802553202.

Bialik, Kristen. 2016. "Most Americans Favor Stricter Environmental Laws and Regulations." *Pew Research Center: FactTank News in the Numbers*. December 14, 2016. Accessed January 29, 2019. https://www.pewresearch.org/fact-tank/2016/12/14/most-americans-favor-stricter-environmental-laws-and-regulations/.

Bike Advisor. 2019. "The State of Bike Commuting in the US: Trends in Prevalence of Bike Commuting Data in the Last Decade." *Bike Advisor*. Accessed April 16, 2022. https://thebikeadviser.com/bike-commuting-united-states/.

Binkley, Sam, and Jo Littler, eds. 2011. *Cultural Studies and Anti-Consumerism: A Critical Encounter*. New York: Routledge.

Bissing-Olson, Megan J., Kelly S. Fielding, and Aarti Lyer. 2016. "Experiences of Pride, Not Guilt, Predict Pro-Environmental Behavior When Pro-Environmental Descriptive Norms Are More Positive." *Journal of Environmental Psychology* 45: 145–53.

Black, Iain. 2020. "Sustainability through Anti-Consumption." *Journal of Consumer Culture Behavior* 9 (6): 403–11. https://doi.org/10.1002/cb.340.

Blake, James. 1999. "Overcoming the 'Value-Action Gap' in Environmental Policy: Tensions between National Policy and Local Experience." *Local Environment* (4) 3: 257–78.

Bonham, Jennifer. 2006. "Transport: Disciplining the Body That Travels." *Editorial Board of the Sociological Review* 54 (1): 57–74. https://doi.org/10.1111/j.1467-954X.2006.00637.x.

Braun, Jennifer, Mary Beckie, and Ken Caine. 2020. "'Trust Us, We Feed This to Our Kids': Women and Public Trust in the Canadian Agri-Food System." *Agriculture and Human Values* 37 (2): 495–507. https://doi.org/10.1007/s10460-019-10002-x.

Brewster, Bradley H., and Michael Mayerfeld Bell. 2010. "The Environmental Goffman: Toward an Environmental Sociology of Everyday Life." *Society and Natural Resources* 23: 45–57. https://www.researchgate.net/publication/232929708.

Breyer, Christian, Sirkka Heinonon, and Juho Ruotsalainen 2017. "New Consciousness: A Societal and Energetic Vision for Rebalancing Humankind within the Limits of Planet Earth." *Technological Forecasting and Social Change* 114: 7–15.

Brady, Emily, and Pauline Phemister, eds. 2012. *Human-Environment Relations: Transformative Values in Theory and Practice.* New York: Springer: 135–48.

Brody, Samuel D., and Hyung-Cheal Ryu. 2006. "Measuring the Educational Impacts of a Graduate Course on Sustainable Development." *Environmental Education Research* 12 (2): 179–99. https://doi.org/10.1080/13504620600688955.

Broom, Catherine. 2017. "Exploring the Relations Between Childhood Experiences in Nature and Young Adults' Environmental Attitudes and Behaviors." *Australian Journal of Environmental Education* 33 (1): 34–47. https://doi.org/10.1017/aee.2017.1

Brown, Kirk Warren, and Tim Kasser. 2005. "Are Psychological and Ecological Well-Being Compatible? The Role of Values, Mindfulness, and Lifestyle." *Social Indicators Research* 74: 349–68.

Buckingham, Susan, and Rakibe Kulcur. 2009. "Gendered Geographies of Environmental Justice." *Antipode* 41 (4): 659–83. https://www.researchgate.net/publication/229738945.

Bunting, Madeleine. 2005. *Willing Slaves: How the Overwork Culture is Ruling Our Lives.* New York: Harper Collins.

Burd, Charlynn, Michael Burrows, and Brian M. Kenzie. 2021. "Travel Time to Work in the United States: 2019." *American Community Survey Reports. United States Census Bureau.* Accessed on July 7, 2022. https://www.census.gov/content/dam/Census/library/publications/2021/acs/acs-47.pdf

Burk, Derek. 2017. "Infrastructure, Social Practice, and Environmentalism: The Case of Bicycle-Commuting." *Social Forces* 95 (3): 1209–36. https://www.jstor.org/stable/26166869.

Burningham, Kate, and Susan Venn. 2020. "Are Lifecourse Transitions Opportunities for Moving to More Sustainable Consumption?" *Journal of Consumer Culture* 20 (1): 102–21. https://doi.org/10.1177/1469540517729010.

Burningham, Kate, Susan Venn, Ian Christie, Tim Jackson, and Birgitta Gatersleben. 2014. "New Motherhood: A Moment of Change in Everyday Shopping Practices." *Young Consumers* 15 (3): 211–26. https://www.proquest.com/docview/1663680693.

Burrows, Michael. 2019. "Younger Workers in Cities More Likely to Bike to Work." *United States Census Bureau*. Accessed on April 16, 2022. https://www.census.gov

/library/stories/2019/05/younger-workers-in-cities-more-likely-to-bike-to-work.html#:~:text=In%202019%2C%20the%20U.S.%20Census,up%20from%20665%2C000%20in%202007.

Cairns, Kate, and Josee Johnston. 2018. "On (Not) Knowing Where Your Food Comes From: Meat, Mothering and Ethical Eating." *Agriculture and Human Values* 35: 569–80. https://doi.org/10.1007/s10460-018-9849-5

Cairns, Kate, Josee Johnston, and Norah MacKendrick. 2013. "Feeding the 'Organic Child': Mothering through Ethical Consumption." *Journal of Consumer Culture* 13 (2): 97–118. https://doi.org/10.1177/1469540513480162.

Carfagna, Lindsey B., Emile A. Dubois, Connor Fitzmaurice, Monique Y. Ouimette, Juliet B. Schor, and Margaret Willis. 2014. "An Emerging Eco-Habitus: The Reconfiguration of High Cultural Capital Practices among Ethical Consumers." *Journal of Consumer Culture* 14 (2): 158–78.

Carolan, Michael S. 2009. "I Do Therefore There Is: Enlivening Socio-Environmental Theory." *Environmental Politics* 18 (1): 1–17. https://doi.org/10.1080/09644010802622748.

Carolan, Michael S. 2011. *Embodied Food Politics*. Farnham, UK: Ashgate Publishing.

CCED (Charleston County Economic Development). 2019. "Climate Data." *Data Center.* Accessed June 5, 2019. http://www.charlestoncountydevelopment.org/data-center/climate-data/.

Chancel, Lucas, and Thomas Piketty. 2015. *Carbon and Inequality: From Tokyo to Paris*. Paris School of Economics 48. Working Paper. Accessed June 21, 2022. https://wid.world/document/chancel-l-piketty-t-carbon-and-inequality-from-kyoto-to-paris-wid-world-working-paper-2015-7/.

Charles, Cheryl, Karen Keenleyside, Rosalie Chapple, Bill Kilburn, Pascale Salah van der Leest, Diana Allen, Miles Richardson, Matteo Giusti, Lawrence Franklin, Michael Harbrow, Ruth Wilson, Andrew Moss, Louise Metcalf, and Luis Camargo. 2018. *Home to Us All: How Connecting with Nature Helps Us Care for Ourselves and the Earth.* Minneapolis: Children & Nature Network. Accessed on June 29, 2022. https://derby.openrepository.com/bitstream/handle/10545/623158/HometoUsAll.pdf?sequence=1

Charleston Moves website. Accessed on March 23, 2022. https://charlestonmoves.org/battery2beach/

Chawla, Louise. 1999. "Life Paths Into Effective Environmental Action." *The Journal of Environmental Education* 31 (1): 15–26.

Chawla, Louise. 2001. "Significant Life Experiences Revisited Once Again: Response to Vol. 5(4) 'Five Critical Commentaries on Significant Life Experience Research in Environmental Education.'" *Environmental Education Research* 7 (4): 451–61.

Chawla, Louise. 2008. "Participation and the Ecology of Environmental Awareness and Action." In *Participation and Learning*, edited by A. Reid, B. B. Jensen, J. Nikel, and V. Simovska, 98–110. New York: Springer.

Chawla, Louise. 2020. "Childhood Nature Connection and Constructive Hope: A Review of the Literature on Connecting with Nature and Coping with Environmental Loss." *British Ecological Society* 2 (3): 619–42.

Chawla, Louise, 2021. "Knowing Nature in Childhood: Learning and Well-Being Through Engagement with the Natural World." In *Nature and Psychology* 67, edited by A. R. Schutte, J. C. Torquati, and J. R. Stevens, 153–93.

Chawla, Louise, and Debra Flanders Cushing. 2007. "Education for Strategic Environmental Behavior." *Environmental Education Research* 13 (4): 437–52.

Chawla, Louise, and Victoria Derr. 2012. "The Development of Conservation Behaviors in Childhood and Youth." *The Oxford Handbook of Environmental and Conservation Psychology,* 527–55. https://doi.org/10.1080/13504620701581539.

Chitewere, Tendai. 2018. *Sustainable Communities and Green Lifestyles: Consumption and Environmentalism*. New York: Routledge.

Claus, Julie M., Mark A. Dessauer, and Laura K. Brennan. 2012. "Programs and Promotions: Approaches by 25 Active Living by Design Partnerships." *American Journal of Preventive Medicine* 43 (5): 320–28.

Clover, Darlene. 2002. "Traversing the Gap: Concientizacion, Educative-Activism in Environmental Adult Education." *Environmental Education Research* 8 (3): 315–23.

Cohen, Maurice, J., Halina Szejnwald Brown, and Philip J. Vergragt, eds. 2013. *Innovations in Sustainable Consumption: New Economic, Socio-technical Transitions and Social Practices.* Northampton: Edward Elgar Publishers.

Colby, Ashley. 2020. "'It connects me to the Earth': Marginalized Environmentalism and a Resistance to Capitalist Logic Among Subsistence Food Producers in Chicago." *Environmental Sociology* 6 (1): 82–94. https://doi.org/10.1080/23251042.2019.1672616.

Collado, Silvia, and Jose A. Corraliza. 2015. "Children's Restorative Experiences and Self-Reported Environmental Behaviors." *Environment and Behavior* 47(1): 38–56. https://doi.org/10.1177/0013916513492417.

Collins, Rebecca. 2015. "Keeping it in the Family? Re-focusing Household Sustainability." *Geoforum* 60: 22–32.

Colville-Andersen, Mikael. 2018. *Copenhagenize: The Definitive Guide to Global Bicycle Urbanism.* Washington, DC: Island Press.

Conroy, Pat. 1986. *The Prince of Tides.* New York: Houghton Mifflin.

Corral-Verdugo, Victor, Jose Mireles-Acosta, Cesar Tapia-Fonllem, and Blanca Fraijo-Sing. 2011. "Happiness as Correlate of Sustainable Behavior: A Study of Pro-Ecological, Frugal, Equitable and Altruistic Actions That Promote Subjective Wellbeing." *Research in Human Ecology* 18 (2): 95–104.

Cousins, Elicia M. 2021. "Mobilizing Motherhood: The Gendered Burden of Environmental Protection." *Sociology Compass* 15 (5): e12879. Accessed on March 20, 2022. https://doi.org/10.1111/soc4.12879.

Cox, Stan, and Paul Cox. 2016. *How the World Breaks: Life in Catastrophe's Path, from the Caribbean to Siberia*. New York: New Press.

Craig, Keltie. 2008. "Red, Black & Shiny: Building Bike Culture in the Urban Environment." A Project Submitted in Partial Fulfillment of the Requirements for the Degree of Master of Arts (Planning). The University of British of Columbia. Accessed on July 5, 2022. file:///C:/Users/Mccarthyd/Downloads/SCARP_May2008_gradproject_Craig.pdf

Cresswell, Timothy. 2006. *On the Move: Mobility in the Modern Western World*. London: Routledge.

Crippa, Monica, Diego Guizzardi, Marilena Muntean, Edwin Solazzo, Efision Monforti-Ferrario, Jos Olivier, and Elisabetta Vignati. 2020. "Fossil CO2 Emissions of All World Countries—2020 Report." *European Commission*. Accessed March 1, 2022. https://publications.jrc.ec.europa.eu/repository/handle/JRC121460.

Crighton, E. J., C. Brown, J. Baxter, L. Lemyre, J. R. Masuda, and F. Ursitti. 2013. "Perceptions and Experiences of Environmental Health Risks among New Mothers: A Qualitative Study in Ontario, Canada." 2013. *Health Risk & Society* 15 (4): 295–312. Accessed on July 2, 2022. https://www.tandfonline.com/doi/pdf/10.1080/13698575.2013.796345?needAccess=true

Crogan, Patrick, and Samuel Kinsley. 2012. "Paying Attention: Toward a Critique of the Attention Economy." *Culture Machine* 13: 1–29. https://uwe-repository.worktribe.com/OutputFile/945732.

Crutzen, Paul, and Eugene Stoermer. 2000. "The Anthropocene." *IGBP Newsletter* 41: 17–18. http://www.igbp.net/globalchange/anthropocene.4.1b8ae20512db692f2a680009238.html.

D'Amore, Chiara. 2016. "Family Nature Clubs: An Intergenerational Opportunity to Foster Love of the Natural World." *Families, Relationships and Societies* 5 (3): 431–46. https://doi.org/10.1332/204674316X14758471234542.

Darlington, Abigail. 2018. "Florence Could Flood Charleston Area More than Tropical Storm Irma Did." *Post and Courier*. September 12, 2018. Accessed January 2, 2019. https://www.postandcourier.com/news/florence-could-flood-charleston-area-more-than-tropical-storm-irma/article_c46c8888-b69a-11e8-b972-4346be6b14b7.html.

Darnton, Andrew, Bas Verplanken, Paul White, and Lorraine Whitmarsh. 2011. "Habits, Routines and Sustainable Lifestyles." *AD Research & Analysis for Defra*. London.

Dehghanmongabadi, Abolfazi, and Hoskara Sebnem. 2020. "An Integrated Framework for Planning Successful Complete Streets: Determinative Variables and Main Steps." *International Journal of Sustainable Transportation* 16 (2): 181–94. https://doi.org/10.1080/15568318.2020.1858373

De Young, Raymond. 2000. "Expanding and Evaluating Motives for Environmentally Responsible Behavior." *Journal of Social Issues* 56 (3): 509–26.

Dias, Maria Accioly, Carlos Frederico B. Loureiro, Leandro Chevitarese, and Cecilia De Mello E. Souza. 2017. "The Meaning and Relevance of Ecovillages for the Construction of Sustainable Alternatives." *Ambiente & Sociedade* 20 (3): 79–96. https://www.academia.edu/35492452/.

Dietz, Thomas, Gerald T. Gardner, Jonathan Gilligan, Paul C. Stern, and Michael P. Vandenbergh. 2009. "Household Actions Can Provide a Behavioral Wedge to Rapidly Reduce U.S. Carbon Emissions." *Proceedings of the National Academy of Sciences* 106 (44): 18452–56. https://doi.org/10.1073/pnas.0908738106.

Dille, Ian. 2016. "The Worst City for Cyclists of 2016: It Seems Like Everyone Loves Bikes in Charleston, South Carolina—Except for the Local Government." *Bicycling: Bikes & Gear Training News*. October 5, 2016. Accessed on June 5,

2019. https://www.bicycling.com/news/a20004709/the-worst-city-for-cyclists-of-2016/.

Diprose, Rosalyn. 1994. *The Bodies of Women: Ethics, Embodiment, and Sexual Differences*. London: Routledge.

Dobernig, Karin, and Sigrid Stagl. 2015. "Growing a Lifestyle Movement? Exploring Identity-Work and Lifestyle Politics in Urban Food Cultivation." *International Journal of Consumer Studies* 39 (5): 452–58. https://doi.org/10.1111/ijcs.12222.

Dombroski, Kelly. 2016. "Hybrid Activist Collectives: Reframing Mothers' Environmental and Caring Labour." *International Journal of Sociology and Social Policy* 36 (9): 629–46. https://www.researchgate.net/publication/309528625.

Donahue, Lillian. 2020. "Wetland Concerns for Long Savannah Project: Environmentalists Say Could Increase Flooding." *Live 5 WCSC News*. August 2, 2022. Accessed on March 22, 2022. https://www.live5news.com/2020/08/01/activist-groups-push-back-long-savannah-project-citing-wetland-development/.

Donald, Betsy, Meric Gertler, Mia Gray, and Linda Lobao. 2010. "Re-regionalizing the Food System?" *Cambridge Journal of Regions, Economy and Society* 3 (2): 171–75.

Dopko, Raelyne L., Colin A. Capaldi, and John M. Zelenski. 2019. "The Psychological and Social Benefits of a Nature Experience for Children: A Preliminary Investigation." *Journal of Environmental Pscychology* 63: 134–38.

Doran, Peter. 2017. *A Political Economy of Attention, Mindfulness and Consumerism: Reclaiming the Mindful Commons.* London: Routledge.

Duron-Ramos, Maria Fernanda, Silvia Collado, Fernanda Inez Garcia-Vazquez, and Maria Bell-Echeverria. 2020. "The Role of Urban/Rural Environments on Mexican Children's Connection to Nature and Pro-Environmental Behavior." *Frontiers in Psychology* 11: 1–6. https://doi.org/10.3389/fpsyg.2020.00514.

Durt, Christoph, Thomas Fuchs, and Christian Tewes. 2017. *Embodiment, Enaction, and Culture: Investigating the Constitution of the Shared World*. Cambridge: The MIT Press.

Eicher-Catt, Deborah. 2001. "The Body as Sign and Embodiment." *The American Journal of Semiotics* 17 (4): 93–130.

Elliott, Rebecca. 2013. "The Taste for Green: The Possibilities and Dynamics of Status Differentiation through 'Green' Consumption." *Poetics* 41 (3): 294–322.

Elvik, Rune, and Rahul Goel. 2019. "Safety-In-Numbers: An Updated Meta-Analysis of Estimates." *Accident Analysis & Prevention* 129: 136–46.

Enzler, Heidi Bruderer. 2013. "Consideration of Future Consequences as a Predictor of Environmentally Responsible Behavior: Evidence from a General Population Study." *Environment and Behavior* 47 (6): 618–43. https://doi.org/10.1177/0013916513512204.

Erlandson, David A., Edward L. Harris, Barbara L. Skipper, and Steve D. Allen. 1993. *Doing Naturalistic Inquiry: A Guide to Methods*. Newbury Park, CA: Sage.

Evans, David, and Wokje Abrahamse. 2009. "Beyond Rhetoric: The Possibilities of and for 'Sustainable Lifestyles.'" *Environmental Politics* 18 (4): 486–502.

Evans, Gary, W. Otto Siegmar, and Florian G. Kaiser. 2018. "Childhood Origins of Young Adult Environmental Behavior." *Psychological Science* 29 (5): 679–87.

Featherstone, Mike, Mike Hepworth, and Bryan S. Turner, eds. 1991. *The Body: Social Process and Cultural Theory.* Thousand Oaks: Sage Publications.

Fernandez-Llamazares, Alvaro, Maria Garteizgogeascoa, Niladri Basu, Eduardo Sonnewend Brondizio, Mar Cabeza, Joan Martinez-Alier, Pamela McElwee, and Victoria Reyes-Garcia. 2020. "A State-of-the-Art Review of Indigenous Peoples and Environmental Pollution." *Integrated Environmental Assessment and Management* 16 (3): 324–41. https://doi.org/10.1002/ieam.4239.

Fincham, Ben. 2006. "Bicycle Messengers and the Road to Freedom." *Sociological Review* 54 (1): 208–22.

Fincham, Ben. 2007. "'Generally Speaking People Are in It for the Cycling and the Beer': Bicycle Couriers, Subculture, and Enjoyment." *Sociological Review* 55 (2): 189–202. https://doi.org/10.1111/j.1467-954X.2007.00701.x.

Flint, Kate. 2001. "Institutional Ecological Footprint Analysis: A Case Study of the University of Newcastle, Australia." *International Journal of Sustainability in Higher Education* 2 (1): 48–62.

Foucault, Michel. 1979 (2012). *Discipline and Punish.* Translated by Alan Sheridan. New York: Vintage Books.

Fox, Karen M., Barbara Humberstone, and Michael Dubnewick. 2014. "Cycling into Sensoria: Embodiment, Leisure, and Tourism." *Tourism Review International* 15 (1–2): 71–85.

Francis, Pope. 2015. "Encyclical Letter: 'Laudato Si': On Care for our Common Home." Huntington: Our Sunday Visitor. Accessed June 3, 2019. http://w2.vatican.va/content/francesco/en/encyclicals/documents/papa-francesco_20150524_enciclica-laudato-si.html

Fredericks, Sarah, E. 2014. "Online Confessions of Eco-Guilt." *Journal for the Study of Religion, Nature, and Culture* 8 (1): 64–84.

Frezza, Marcia, Lorraine Whitmarsh, Martina Shafer, and Ulf Schrader. 2019. "Spillover Effects of Sustainable Consumption: Combining Identity Process Theory and Theories of Practice." *Sustainability, Science, Practice and Policy* 15 (1): 15–30.

Fuchs, Doris, Antonietta Di Giulio, Katharina Glaab, Sylvia Lorek, Michael Maniates, Thomas Princen, and Inge Ropke. 2016. "Power: The Missing Element in Sustainable Consumption and Absolute Reductions Research and Action." *Journal of Cleaner Production* 132: 298–307. https://doi.org/10.1016/j.jclepro.2015.02.006.

Gale, Jen. 2020. *The Sustainable(ish) Guide to Green Parenting: Guilt-Free Eco-Ideas for Raising Your Kids.* Dublin, Ireland: Bloomsbury, Green Tree.

Gao, Zhihong. 2018. "Mapping the Official Discourse of Frugality in China between 1979 and 2015." *Journal of Historical Research in Marketing* 10 (2): 151–74.

Geels, Frank W. 2011. "The Multi-level Perspective on Sustainability Transitions: Responses to Seven Criticisms." *Environmental Innovation and Societal Transitions* 1 (1): 24–40. https://doi.org/10.1016/j.eist.2011.02.002.

Geisler, Charles, and Ben Currens. 2017. "Impediments to Inland Resettlement Under Conditions of Accelerated Sea Level Rise." *Land Use Policy* 66: 322–30. https://doi.org/10.1016/j.landusepol.2017.03.029.

Genus, Audley, and Anne-Marie Coles. 2008. "Rethinking the Multi-Level Perspective of Technological Transitions." *Research Policy* 37 (9): 1436–45. https://doi.org/10.1016/j.respol.2008.05.006.

GhaffarianHoseini, AmirHosein. 2012. "The Essence of Malay Vernacular Houses: Towards Understanding the Socio-cultural and Environmental Values." *Journal of the International Society for the Study of Vernacular Settlements* 2 (2): 53–73. Accessed on June 29, 2022. https://isvshome.com/pdf/ISVS_2-2/ISVS-2-2%203%20Amirhosein-Final-Paper.pdf

Gibson, Chris, Lesley Head, Nick Gill, and Gordon Waitt. 2011. "Climate Change and Household Dynamics: Beyond Consumption, Unbounding Sustainability." *Transactions of the Institute of British Geographers, New Series* 36 (1): 3–8.

Gifford, Robert, and Andreas Nilsson. 2014. "Personal and Social Factors that Influence Pro-Environmental Concern and Behavior: A Review." *International Journal of Psychology 49* (3): 141–57.

Giurge, Laura M., Ashley V. Whillans, and Colin West. 2020. "Why Time Poverty Matters for Individuals, Organisations, and Nations." *Nature, Human Behavior* 4, 993–1003. https://doi.org/10.1038/s41562-020-0920-z.

Gkargkavouzi, Anastasia, George Halkos, and Steriani Matsiori. 2018. "A Multi-Dimensional Measure of Environmental Behavior: Exploring the Predictive Power of Connectedness to Nature, Ecological Worldview, and Environmental Concern." *Social Indicators Research* 143: 859–79.

Global Footprint Network. 2017. *Ecological Footprint of Countries 2017.* New York University and Footprint Data Foundation. Accessed on March 15, 2022. https://data.footprintnetwork.org/?_ga=2.213014356.193975080.1607636931-1228825067.1607467693#/compareCountries?cn=all&type=EFCpc&yr=2017.

Godin, Laurence, Senja Laakso, and Marlyne Sahakian. 2020. "Doing Laundry in Consumption Corridors: Wellbeing and Everyday Life." *Sustainability, Science, Practice and Policy* 16 (1): 90–113. https://doi.org/10.1080/15487733.2020.1785095.

Gough, Ian. 2017. *Heat, Greed, and Human Need: Climate Change, Capitalism and Sustainable Wellbeing*. Northampton, Edward Elgar Publishing.

Greenspan, Itay, Tally Katz-Gerro, Femida Handy, and Sharon Park. 2022. "Intergenerational Transmission of Environmental Household Practices Among South Korean Families: Continuity and Change." *Families, Relationships and Societies* 11 (1): 101–25. https://doi.org/10.1332/204674321X16172729233053.

Grin, John, Jan Rotmans, and Johan Schot. 2011. *Transitions to Sustainable Development*. Routledge: London.

Gronhoj, Alice, and John Thogersen. 2012. "Action Speaks Louder than Words: The Effect of Personal Attitudes and Family Norms on Adolescents' Pro-Environmental Behaviour." *Journal of Economic Psychology* 33 (1): 292–302. https://doi.org/10.1016/j.joep.2011.10.001.

Hagmann, David, Emily H. Ho, and George Loewenstein. 2019. "Nudging Out Support for a Carbon Tax." *Nature Climate Change* 9: 484–89.

Hajat, Anjum, Charlene Hsia, and Marie S. O'Neill. 2015. "Socioeconomic Disparities and Air Pollution Exposure: A Global Review." *Current Environmental Health Reports* 2 (4): 440–50. https://doi.org/10.1007/s40572-015-0069-5.

Halani, Hannah Alani. 2018. "More Legal Mexican Farm Workers Are Coming to SC than Ever. But Few Are Protecting Them." *Post and Courier*. October 17, 2018. Last updated September 14, 2020. https://www.postandcourier.com/news/more-legal-mexican-farm-workers-are-coming-to-sc-than-ever-but-few-are-protecting/article_aa25bd20-ccbe-11e8-a53b-0bca74bc4bb0.html.

Halfacre, Angela C. 2012. *A Delicate Balance: Constructing a Conservation Culture in the South Carolina Lowcountry.* Columbia: University of South Carolina Press.

Halkier, Bente, Tally Katz-Gerro, and Lydia Martens. 2011. "Applying Practice Theory to the Study of Consumption: Theoretical and Methodological Considerations." *Journal of Consumer Culture* 11 (1): 3–13. https://doi.org/10.1177/1469540510391765.

Hamilton, Erin Miller, Meaghan L. Guckian, and Raymond De Young. 2018. "Living Well and Living Green: Participant Conceptualizations of Green Citizenship." In *Handbook of Sustainability and Social Science Research*, edited by Walter L. Filho, Robert W. Marans, and John Callewaert, 315–35. World Sustainability Series. New York: Springer International Publishing.

Han, Heesup, Jongsik Yu, Hyeon-Cheol Kim, and Wansoo Kim. 2018. "The Impact of Social/Personal Norms and Willingness to Sacrifice on Young Vacationers' Pro-Environmental Intentions for Waste Reduction and Recycling." *Journal of Sustainable Tourism* 26 (12). Accessed on July 2, 2022. https://doi.org/10.1080/09669582.2018.1538229

Handy, Susan L., Marlon G. Boarnet, Reid Ewing, and Richard E. Killingsworth. 2002. "How the Built Environment Affects Physical Activity: Views from Urban Planning." *American Journal of Preventive Medicine* 23 (2): 64–73. https://doi.org/10.1016/s0749-3797(02)00475-0.

Hards, Sarah. 2011. "Social Practice and the Evolution of Personal Environmental Values." *Environmental Values* 20: 23–42.

Hargreaves, Tom. 2011. "Practice-ing Behaviour Change: Applying Social Practice Theory to Pro-Environmental Behavior Change." *Journal of Consumer Culture* 11 (1):79–99. https://doi.org/10.1177/1469540510390500.

Hargreaves, Tom. 2016. "Interacting for the Environment: Engaging Goffman in Pro-Environmental Action." *Society and Natural Resources* 29 (1): 53–67. https://doi.org/10.1080/08941920.2015.1054978.

Hart, Brodie. 2020. "'This Did Not Have to Happen': Bicyclists Advocate for a Safer Way to Cross Cosgrove Bridge." *ABC News*. May 28, 2020. Accessed on May 27, 2022. https://abcnews4.com/news/local/this-did-not-have-to-happen-bicyclists-advocate-for-safer-way-to-cross-cosgrove-bridge.

Harvey, Deborah J., Louise N. Montgomery, Hannah Harvey, Felix Hall, Alan C. Gange, and Dawn Watling. 2020. "Psychological Benefits of a Biodiversity-Focused Outdoor Learning Program for Primary School Children." *Journal of Environmental Psychology* 67 Article 101381.

Harvey, Andrew S., and Arun K. Mukhopadhyay. 2007. "When Twenty-Four Hours Is Not Enough: Time Poverty of Working Parents." *Social Indicator Research* (82): 57–77. https://doi.org/10.1007/s11205-006-9002-5.

Hasselqvist, Hanna, and Mia Hesselgren. 2019. "Bridging Citizen and Stakeholder Perspectives of Sustainable Mobility through Practice-Oriented Design." *Sustainability: Science, Practice and Policy* 15 (1): 1–14. https://doi.org/10.1080/15487733.2018.1533781.

Haswell, Peter M. 2021. "Scientist Warning on Why You Should Consume Less; Even if Wider Society Doesn't." *Nature and Culture* 16 (3): 29–48. https://doi.org/10.3167/nc.2021.160302.

Hatala, Andrew R., Chinyere Njeze, Darrien Morton, Tamara Peal, and Kelley Bird-Naytowtow. 2020. "Land and Nature as Sources of Health and Resilience among Indigenenous Youth in an Urban Canadian Context: A Photovoice Exploration." *BMC Public Health* 20: 1–4.

Haxeltine, Alex, Flor Avelino, Julia M. Wittmayer, Iris Kunze, Noel Longhurst, Adina Dumitru, and Tim O'Riordan. 2017. "Conceptualizing the Role of Social Innovation in Sustainability Transformations." In *Social Innovation and Sustainable Consumption: Research and Action for Societal Transformation*, edited by Julia Backhaus, Audley Genus, Sylvia Lorek, Edina Vadovics, and Julia M. Wittmayer, 12–25. New York: Routledge.

Hedlund-de Witt, Annick. 2012. "Exploring Worldviews and Their Relationships to Sustainable Lifestyles: Towards a New Conceptual and Methodological Approach." *Ecological Economics* 84: 74–83.

Hedlund-de Witt, Annick. 2014. "The Integrative Worldview and Its Potential for Sustainable Societies: A Qualitative Exploration of the Views and Values of Environmental Leaders." *Worldviews: Global Religions, Culture & Ecology* 18: 191–229. http://www.jstor.org/stable/43809586.

Hedlund-de Witt, Annick, Patricia Osseweijer, and Robin Pierce. 2015. "Understanding Public Perceptions of Biotechnology through the 'Integrative Worldview Framework.'" *Public Understanding of Science* 26 (1): 70–88. Accessed on September 21, 2011. https://doi.org/10.1177%2F0963662515592364

Heimlich, Joe E., and Nicole M. Ardoin. 2008. "Understanding Behavior to Understand Behavior Change: A Literature Review." *Environmental Education Research* 14 (3): 215–37. https://doi.org/10.1080/13504620802148881.

Helne, Tuula, and Tuuli Hirvilammi. 2019. Chapter 13, "Having, Doing, Loving, Being: Sustainable Well-Being for a Post-Growth Society." In *Towards a Political Economy of Degrowth*, edited by E. Chertkovskaya, A. Paulsson, and S. Barca, 225–42. Lanham: Rowman & Littlefield.

Helne, Tuula, and Arto O. Salonen. 2016. "Ecosocial Food Policy: Improving Human, Animal, and Planetary Well-Being." *Sustainability: Science, Practice & Policy* 12 (2): 1–11.

Helne, Tuula, and Tuuli Hirvilammi. 2015. "Wellbeing and Sustainability: A Relational Approach." *Sustainable Development* 23: 167–75.

Hendlin, Yogi Hale. 2016. "Bicycling and the Politics of Recognition." In *The Greening of Everyday Life: Challenging Practices, Imagining Possibilities*, edited by John M. Meyer and Jens M. Kersten, 230–46. Oxford: Oxford University Press.

Henwood, Karen, Chris Groves, and Fiona Shirani. 2016. "Relationality, Entangled Practices, and Psychosocial Exploration of Intergenerational Dynamics in Sustainable Energy Studies." *Families, Relationships, Societies* 5 (3) 393–410.

Hines, Jody M., Harold R. Hungerford, and Audrey N. Tomera. 1987. "Analysis and Synthesis of Research on Responsible Environmental Behavior: A Meta-Analysis." *Journal of Environmental Education* 18 (2): 1–8. https://doi.org/10.1080/00958964.1987.9943482.

Hinkley, Trina, Helen Brown, Valerie Carson, and Megan Teychenne. 2018. "Cross Sectional Associations of Screen Time and Outdoor Play with Social Skills in Preschool Children." *PLOS ONE* 13 (4). Accessed on July 1, 2022. https://doi.org/10.1371/journal.pone.0193700

Hitchings, Russell, Rebecca Collins, and Rosie Day. "Inadvertent Environmentalism and the Action-Value Opportunity: Reflections from Studies at Both Ends of the Generational Spectrum." *Local Environment: The International Journal of Justice and Sustainability* 20 (3): 369–85.

Hochgerner, Josef. 2011. "The Analysis of Social Innovations as Social Practice." In *Die Analyse Socialer Innovationen als gesellschaftliche Praxis*, edited by J. Hochgerner, 173–89. Vienna and Berlin: LIT. Accessed on September 21, 2011. https://www.zsi.at/object/publication/1566/attach/The%20Analysis%20of%20Social%20Innovations%20as%20Social%20Practice.pdf

Hochli, Bettina, Adrian Brugger, Roman Abegglen, and Claude Messner. 2019. "Using a Goal Theoretical Perspective to Reduce Negative and Promote Positive Spillover After a Bike-to-Work Campaign." *Frontiers in Psychology* 10 (433): 1–18. Accessed on June 19, 2022. https://doi.org/10.3389/fpsyg.2019.00433

Hoffmann, Melody, L. 2016. *Bike Lanes Are White Lanes: Bicycle Advocacy and Urban Planning*. Lincoln: University of Nebraska Press.

Horton, Dave, Paul Rosen, and Peter Cox, eds. 2007. *Cycling and Society*. New York: Routledge.

Howell, Rachel A. 2013. "It's not (just) 'the environment stupid!' Values, Motivations, and Routes to Engagement of People Adopting Lower-Carbon Lifestyles." *Global Environmental Change* 23: 281–90.

Huddart Kennedy, Emily, Maurie J. Cohen, and Naomi T. Krogman, eds. 2015. *Putting Sustainability Into Practice: Applications and Advances in Research on Sustainable Consumption*. Northampton: Edward Elgar Publishing.

Huddart Kennedy, Emily, and Julie A. Kmec. 2019. "Is There an 'Ideal Feeder'? How Healthy and Eco-Friendly Food Consumption Choices Impact Judgements of Parents." *Agriculture and Human Values* 36 (1): 137–51. https://doi.org/10.1007/s10460-018-09904-z.

Huddart Kennedy, Emily, and Jennifer E. Givens. 2019. "Eco-Habitus or Eco-Powerlessness? Examining Environmental Concern across Social Class." *Sociological Perspectives* 62 (5): 646–67.

Hull, Angela, and Craig O'Holleran. 2014. "Bicycle Infrastructure: Can Good Design Encourage Cycling?" *Urban Planning and Transport Research* 2 (1): 369–406.

Humberstone, Barbara. 2010. "Engagements with Nature: Ageing and Windsurfing." In *Identities, Cultures, and Voices in Leisure and Sport,* edited by B. Watson and J. Harpin, 159–69. Association Publication No. 116. Eastbourne: Leisure Studies.

Humberston, Barbara. 2011. "Embodiment and Social and Environmental Action in Nature-Based Sport: Spiritual Spaces." *Journal of Leisure Studies* 30 (4): 495–512.

Humberstone, Barbara. 2013. "Adventurous Activities, Embodiment and Nature: Spiritual, Sensual, and Sustainable? Embodying Environmental Justice." *Motriz. Revista de Educacao Fisica* 19 (3): 565–71.

Humberstone, Barbara. 2015. Chapter 6, "Embodiment, Nature and Wellbeing." In *Experiencing the Outdoors: Enhancing Strategies for Wellbeing*, edited by M. Robertson and R. Lawrence. Boston: Brill: 61–72.

Humphrey, Kim. 2009. *Excess: Anti-Consumerism in the West*. New York: Polity Press.

Huntsman, Anna, and Jake Steinberg. 2019. "Developments in Disaster-Prone Areas Mean Big Bucks for Developers but Can Put Homeowners at Risk." *Carnegie-Knight News21*. August 15, 2019. Accessed on March 22, 2022. https://cronkitenews.azpbs.org/2019/08/15/developments-disaster-prone-communities-flooding/.

Hurtado, Petra Stieninger. 2018. "From Sustainable Cities to Sustainable People—Changing Behavior Towards Sustainability with the Five A Planning Approach." In *Handbook of Sustainability and Social Science Research*, edited by W. Filho, R. Marans, and J. Callewaert. New York: Springer International Publishing.

Ingold, Tim. 2004. "Culture on the Ground: The World Perceived Through Feet." *Journal of Material Culture* 9 (3): 315–40. https://doi.org/10.1177/1359183504046896.

Ingold, Tim. 2011a. *Being Alive: Essays on Movement, Knowledge and Description*. Oxfordshire, UK: Routledge.

Ingold, Tim. 2011b/2000. T*he Perception of the Environment: Essays on Livelihood, Dwelling and Skill*. Oxfordshire, UK: Routledge.

Intergovernmental Panel on Climate Change (IPCC). 2022. *Climate Change 2022: Impacts, Adaptation, and Vulnerability*. Working Group II Contribution to the IPCC Sixth Assessment Report. Accessed on March 15, 2022. https://www.ipcc.ch/report/sixth-assessment-report-working-group-ii/.

Ivanova, Diana, Konstantin Stadler, Kjartan Steen-Olsen, Richard Wood, Gibran Vita, Arnold Tukker, and Edgar G. Hertwich. 2016. "Environmental Impact Assessment of Household Consumption." *Journal of Industrial Ecology* 20 (3): 526–36. https://doi.org/10.1111/jiec.12371.

Jack, Tullia. 2013. "Nobody Was Dirty: Intervening in Inconspicuous Consumption of Laundry Routines." *Journal of Consumer Culture* 13 (3): 406–21. https://doi.org/10.1177/1469540513485272.

Jackson, Jeffrey B., Michael X. Kirby, Wolfgang H. Berger, Karen A. Bjorndal, Louis W. Botsford, Bruce J. Bourque, Roger H. Bradbury, . . . and Robert R. Warner.

2001. "Historical Overfishing and the Recent Collapse of Coastal Ecosystems." *Science* 293 (5530): 629–37. https://doi.org/10.1126/science.1059199.

Jaeger-Erben, Melanie, and Jana Ruckart-John. 2015. "Researching Transitions to Sustainable Consumption: A Practice Theory Approach to Studying Innovations in Consumption." In *Putting Sustainability Into Practice: Applications and Advances in Research on Sustainable Consumption*, edited by Emily Huddart Kennedy, Maurie J. Cohen, and Naomi T. Krogman, 159–84. Northampton: Edward Elgar Publishing.

Jamail, Dahr. 2019. *The End of Ice: Bearing Witness and Finding Meaning in the Path of Climate Change*. New York: New Press.

Jamieson, Lynn. 2016. "Families, Relationships, and 'Environment': (Un)sustainability, Climate Change, and Biodiversity Loss." *Families, Relationships, and Societies* 5 (3): 335–55.

Jarosz, Lucy. 2000. "Understanding Agri-food Networks as Social Relations." *Agriculture and Human Values* 17 (3): 279–83.

Jayaratane, Menuka, Gillian Sullivan Mort, and Clare D'Souza. 2015. "Sustainability Living in a Carbon-Priced Economy: 'Shoulds' and 'Woulds," Making Amends and Sustainability Guilt." *Journal of Nonprofit and Public Sector Marketing* 27 (3): 285–306.

Jensen, Bjarne Bruun. 2002. "Knowledge, Action, and Pro-Environmental Behaviour." *Environmental Education Research* 8 (3): 325–34. https://doi.org/10.1080/13504620220145474.

Jones, Phil. 2005. "Performing the City: A Body and a Bicycle Take on Birmingham, UK." *Social and Cultural Geography* 6 (6): 813–30. https://doi.org/10.1080/14649360500353046.

Kahneman, Daniel, Edward Diener, and Ed Diener, eds. 1999. *Well-being: The Foundations of Hedonic Psychology*. New York: Russell Sage Foundation.

Kalbar, Pradip P., Morten Birkved, Michael Hauschild, Simon Kabins, and Simon Elsborg Nygaard. 2018. "Environmental Impact of Urban Consumption Patterns: Drivers and Focus Points." *Resources, Conservation and Recycling* 137: 260–69. https://doi.org/10.1016/j.resconrec.2018.06.019.

Kalof, Linda, Thomas Dietz, Gregory Guagnano, and Paul C. Stern. "Race, Gender, and Environmentalism: The Atypical Values and Beliefs of White Men." *Race, Gender, and Class* 9 (2): 112–30.

Karlsson, Rasmus. 2012. "Individual Guilt or Collective Progressive Action? Challenging the Strategic Potential of Environmental Citizenship Theory." *Environmental Values* 21 (4): 459–74.

Kelley, Rebecca, Joy Hatch, and Josh Dorfman. 2010. *The Eco-nomical Baby Guide: Down-to-Earth Ways for Parents to Save Money and the Planet*. New York: Stewart, Tabori, and Chang.

Kennedy, Emily Hubbart, Harvey Krahn, and Naomi T. Krogman. 2015. "Are We Counting What Counts? A Closer Look at Environmental Concern, Pro-Environmental Behaviour, and Carbon Footprint." *Local Environment* 20 (2): 220–36. https://doi.org/10.1080/13549839.2013.837039.

Kidder, Jeffrey L. 2005. "Style and Action: A Decoding of Bike Messenger Symbols." *Journal of Contemporary Ethnography* 34 (3): 344–67. https://doi.org/10.1177/0891241605274734.

Kidder, Jeffrey L. 2006a. "Bike Messengers and the Really Real: Effervescence, Reflexivity, and Postmodern Identity." *Symbolic Interaction* 29 (3): 349–71. http://commons.lib.niu.edu/handle/10843/13344.

Kidder, Jeffrey L. 2006b. "'It's the Job That I Love': Bike Messengers and Edgework." *Sociological Forum* 21 (1): 31–54. https://www.jstor.org/stable/4540926.

Kidder, Jeffrey L. 2008. "Appropriating the City: Space, Theory, and Bike Messengers." *Theory and Society* 38 (3): 307–28. https://doi.org/10.1007/s11186-008-9079-8.

Kidder, Jeffrey L. 2011. *Urban Flow: Bike Messengers and the City*. Ithaca, NY: Cornell University Press.

Kirby, Andy. 2003. "Redefining Social and Environmental Relations at the Ecovillage at Ithaca: A Case Study." *Journal of Environmental Psychology* 23: 323–32.

Kofsky, Jared, and Cameron Bopp. 2021. "'Transit Deserts' Remain as Charleston Continues to Grow." Live 5 News. https://www.live5news.com/2021/12/14/transit-deserts-remain-charleston-continues-grow/

Kolbert, Elizabeth. 2015. *The Sixth Extinction: An Unnatural History*. New York: Picador.

Kollmus, Anja, and Julian Agyeman. 2002. "Mind the Gap: Why Do People Act Environmentally and What Are the Barriers to Pro-Environmental Behavior?" *Environmental Education* 8 (3): 239–60.

Knez, Igor. 2016. "Is Climate Change a Moral Issue? Effects of Egoism and Altruism on Pro-Environmental Behavior." *Current Urban Studies* 4 (2): 157–74.

Krepelkova, Sarka Dolezalova, Jan Krajhanzi, and Roman Kroufek. 2020. "The Influence of Interaction with Nature in Childhood on Future Pro-Environmental Behavior." *Journal of Baltic Science Education* 19 (4): 536–50. https://eric.ed.gov/?id=EJ1264569.

Kuo, Ming, Michael Barnes, and Catherine Jordan. 2019. "Do Experiences with Nature Promote Learning? Converging Evidence of a Cause-and-Effect Relationship." *Frontiers in Psychology* 10. Article 305. Accessed on June 1, 2022. https://doi.org/10.3389/fpsyg.2019.00305

Larsen, Jonas. 2018. "Commuting, Exercise, and Sport: An Ethnography of Long-Distance Bike Commuting." *Social and Cultural Geography* 19 (1): 39–58. https://doi.org/10.1080/14649365.2016.1249399.

Larson, Lincoln R., Rachel Szczytko, Edmond P. Bowers, Lauren E. Stephens, Kathryn T. Stevenson, and Myron F. Floyd. 2019. "Outdoor Time, Screen Time, and Connection to Nature: Troubling Trends Among Rural Youth?" *Environment and Behavior* 5 (8): 966–91.

Lee, Do Jun. 2015. "Embodied Bicycle Commuters in a Car World." *Social and Cultural Geography* 17(3): 401–22. https://doi.org/10.1080/14649365.2015.1077265.

Lestar, Tamas, and Steffen Bohm. 2020. "Ecospirituality and Sustainability Transitions: Agency Towards Degrowth." *Religion, State and Society* 48 (1): 56–73. https://doi .org/10.1080/09637494.2019.1702410.

Liftin, Karen. 2016. "Ontologies of Sustainability in Ecovillage Culture." In *The Greening of Everyday Life: Challenging Practices, Imagining Possibilities*, edited by John M. Meyer and Jens M. Kersten, 249–64. Oxford: Oxford University Press.

Lincoln, Yvonna, S. and Egon G. Guba. 1985. *Naturalistic Inquiry.* Thousand Oaks: Sage Publishing.

Liu, Michelle. 2020. "Charleston Development Will Worsen Flooding, Challengers Say." *AP News*. June 10, 2020. Accessed on March 22, 2022. https://apnews.com/ article/environment-floods-south-carolina-wildlife-wetlands-

Loreau, Michel, Bradley J. Cardinale, Forest Isbell, Tim Newbold, Mary I. O'Connor, and Claire de Mazancourt. 2022. "Do Not Downplay Biodiversity Loss." *Nature* 601, E27–E28. Accessed on March 19, 2022. https://www.nature.com/articles/ s41586-021-04179-7.

Lorenzen, Janet A. 2012. "Going Green: The Process of Lifestyle Change." *Sociological Forum* 27 (1): 94–116. https://doi.org/10.1111/j.1573-7861.2011 .01303.x.

Lorenzen, Janet A. 2014a. "Convincing People to Go Green: Managing Strategic Action by Minimizing Political Talk." *Environmental Politics* 23 (3): 454–72. https://doi.org/10.1080/09644016.2013.859778.

Lorenzen, Janet A. 2014b. "Green Consumption and Social Change: Debates Over Responsibility, Private Action, and Access." *Sociology Compass* 8 (8): 1063–81. https://doi.org/10.1111/soc4.12198.

Louv, Richard. 2008. *Last Child in the Woods: Saving Our Children from Nature-Deficit Disorder.* Chapel Hill, NC: Algonquin Books.

Lubitow, Amy, Bryan Zinschlag, and Nathan Rochester. 2106. "Plans for Pavement or for People? The Politics of Bike Lanes on the 'Paseo Boricua' in Chicago, Illinois." *Urban Studies* 53 (12): 2637–53. https://doi.org/10.1177/0042098015592823.

Macias, Thomas, and Kristin Williams. 2014. "Know Your Neighbors, Save the Planet: Social Capital and the Widening Wedge of Pro-Environmental Outcomes." *Environment and Behavior* 43 (3): 391–420.

MacKendrick, Norah A. 2010. "Media Framing of Body Burdens: Precautionary Consumption and the Individualization of Risk." *Sociological Inquiry* 80 (1): 126–49. https://doi.org/10.1111/j.1475-682X.2009.00319.x.

MacKendrick, Norah. 2014. "More Work for Mother: Chemical Body Burdens as Maternal Responsibility." *Gender and Society* 28 (5): 705–28. https://doi.org/10 .1177/0891243214529842.

MacKendrick, Norah. 2018. *Better Safe than Sorry: How Consumers Navigate Exposure to Everyday Toxics*. Oakland, CA: University of California Press.

MacKendrick, Norah, and Kate Cairns. 2019. "The Polluted Child and Maternal Responsibility in the US Environmental Movement." *Signs* 44 (2): 307–32. https: //doi.org/10.1086/699340.

MacKendrick, Norah, and Lindsey M. Stevens. 2016. "'Taking Back a Little Bit of Control': Managing the Contaminated Body Through Consumption." *Sociological Forum* 31 (2): 310–29. https://doi.org/10.1111/socf.12245.

Macnaghten, Phil, and John Urry. 2000a. "Bodies of Nature: Introduction." *Body & Society* 6 (3/4): 1–11. https://doi.org/10.1177/1357034X00006003001

Macnaghten, Phil, and John Urry. 2000b. "Bodies in the Woods." *Body & Society* 6 (3–4): 166–82. https://doi.org/10.1177/1357034X00006003009.

Mahdawi, Arwa. 2020. "Most Plastic Will Never Be Recycled—and the Manufacturers Couldn't Care Less." *Guardian*. September 15, 2020. Accessed June 12, 2022. https://www.theguardian.com/commentisfree/2020/sep/15/most-plastic-will-never-be-recycled-and-the-manufacturers-couldnt-care-less.

Maniates, Michael. 2002. "Individualization: Plant a Tree, Buy a Bike, Save the World?" In *Confronting Consumption*, edited by Thomas Princen, Michael Maniates, and Ken Conca, 43–66. Cambridge, MA: MIT Press.

Marcus, Ellen, and Sander Marcus. 2013. *The Fish On Your Dish.* N.p.: Ellen Marcus.

Marglin, Stephen A. and, Juliet B. Schor, eds. 1991. *The Golden Age of Capitalism: Reinterpreting the Postwar Experience.* Oxford: Oxford University Press.

Marinho, Alcyane, and Heloisa Turini-Bruhns. 2005. "Body Relationships in an Urban Adventure Setting." *Leisure Studies* 24 (3): 223–38. https://doi.org/10.1080/0261436052000327234.

Markard, Jochen, Rob Raven, and Bernhard Truffer. 2012. "Sustainability Transitions: An Emerging Field of Research and Its Prospects." *Research Policy* 41 (6): 955–67. https://doi.org/10.1016/j.respol.2012.02.013.

Marshall, Wesley E., and Norman W. Garrick. 2011. "Evidence on Why Bike-Friendly Cities are Safer for All Road Users." *Environmental Practice* 13 (1): 16–27.

Marshall, Wesley E., and Nicholas N. Ferenchak. 2019. "Why Cities with High Bicycling Rates are Safer for All Road Users." *Journal of Transport & Health* 13: 285–301.

Martin, Emanuel. 2019. "Making a Bicycle City: Infrastructure and Cycling in Copenhagen Since 1880." *Urban History* 46 (3): 493–517. https://doi.org/10.1017/S0963926818000573.

Matusik, Jan, and Vladimir Koci. 2019. "Environmental Impact of Personal Consumption from Life Cycle Perspective: A Czech Republic Case Study." *Science of the Total Environment* 646 (1): 177–86. https://doi.org/10.1016/j.scitotenv.2018.07.233.

Maye, Damien, Lewis Holloway, and Moya Kneafsey. 2007. *Alternative Food Geographies: Representation and Practice*. New York: Elsevier.

Mayer, Stephan F., and Cynthia McPherson Frantz. 2004. "The Connectedness to Nature Scale: A Measure of Individuals' Feeling in Community with Nature." *Journal of Environmental Psychology* 24 (4): 503–15.

Mazar, Nina, and Chen-Bo Zhong. 2010. "Do Green Products Make Us Better People?" *Psychological Science* 21 (4): 494–98. https://doi.org/10.1177/0956797610363538.

McCarthy, Deborah. 2010. "Perceptions about and Use of a New Scenic Bridge Path Among Walkers, Runners, and Cyclists in Lowcountry South Carolina." *Preventive Medicine* 51 (1): 94–95. https://doi.org/10.1016/j.ypmed.2010.04.007.

McCarthy, Deborah. 2011. "'I'm a Normal Person': An Examination of How Utilitarian Cyclists in Charleston South Carolina Use an Insider/Outsider Framework to Make Sense of Risks." *Urban Studies* 48 (7): 1439–55. https://doi.org/10.1177/0042098010375322.

McKenzie-Mohr, Doug, and William A. Smith. 1999. *Fostering Sustainable Behavior: An Introduction to Community-Based Social Marketing*. Gabriola Island, BC: New Society Publishers.

McKibben, Bill. 2019. *Falter: Has the Human Game Begun to Play Itself Out?* New York: Henry Holt.

McKillop, Matt, Rachel Pittluck, Jonathan M. Links, Megan Weil Latshaw, Crystal R. Watson, and Tara Kirk Sell. 2020. "Climate Change & Health: Assessing State Preparedness, Is This the United Fate of America?" *Trust for America and the Johns Hopkins Bloomberg School of Public Health*. Accessed on March 15, 2022. https://climateandhealthreport.org/.

McMillan, Tracy E. 2007. "The Relative Influence of Urban Form on a Child's Travel Mode to School." *Transportation Research Part A Policy and Practice* 41 (1): 69–79. https://doi.org/10.1016/j.tra.2006.05.011.

Merleau-Ponty, Maurice. 1945/1962/1992. *The Phenomenology of Perception*. New York: Routledge.

Mert-Cakal, Tezcan, and Mara Miele. 2021. "'Workable Utopias' for Social Change through Inclusion and Empowerment? Community-Supported Agriculture (CSA) in Wales as Social Innovation." *Agriculture and Human Values* 37 (4): 1241–60. https://doi.org/10.1007/s10460-020-10141-6.

Messerschmidt, James W. 2004. *Flesh and Blood: Adolescent Gender Diversity and Violence*. Boulder, CO: Westview Press.

Meyer, John M., and Jens M. Kersten, eds. 2016. *The Greening of Everyday Life: Challenging Practices, Imagining Possibilities*. Oxford: Oxford University Press.

Micheletti, Michele, and Dietlind Stolle. 2012. "Sustainable Citizenship and the New Politics of Consumption." *Annals of American Academy of Political and Social Science* 644 (1): 88–120. https://doi.org/10.1177/0002716212454836.

Mirosa, Miranda, Rob Lawson, and Daniel Gnoth. 2013. "Linking Personal Values to Energy-Efficient Behaviors in the Home." *Environment and Behavior* 45 (4): 455–75. https://doi.org/10.1177/0013916511432332.

Mkono, Mucha. 2020. "Eco-Guilt and Eco-Shame in Tourism Consumption Contexts: Understanding the Triggers and Responses." *Journal of Sustainable Tourism* 28 (8): 1223–44.

Mohai, Paul, and Bunyan Bryant. 1998. "Is There a 'Race' Effect on Concern for Environmental Quality?" *Public Opinion Quarterly* 62: 475–505.

Moisander, Johanna, and Sinnika Pesonen. 2002. "Narratives of Sustainable Ways of Living: Constructing the Self and the Other as a Green Consumer." *Management Decision* 40 (4): 329–42.

Molinario, Erica, Caterina Lorenzi, Flavia Bartoccioni, Paola Perucchini, Sebastian Bobeth, and Agathe Colleony. 2020. "From Childhood Nature Experiences to Adult Pro-Environmental Behaviors: An Exploratory Model of Sustainable Food

Consumption." *Environmental Education Research* 26 (8): 1137–63. https://doi.org/10.1080/13504622.2020.1784851.

Montgomery Fate, Tom. 2012. *Cabin Fever: A Suburban Father's Search for the Wild.* Boston: Beacon Press.

Moore, Thad. 2018. "South Carolina's Technology Sector Grew Steadily Last Year, Adding 2,500 Jobs." *Post and Courier*. April 2, 2018. Accessed January 17, 2019. https://www.postandcourier.com/business/south-carolinas-technology-sector-grew-steadily-last-year-adding-2-500-jobs/article_56b6f124-3392-11e8-a752-5b4b0d4ed286.html

Mostafa, Mohamed M. 2013. "Wealth, Post-Materialism and Consumers' Pro-Environmental Intentions: A Multilevel Analysis Across 25 Nations." *Sustainable Development* 21 (6): 385–99. https://doi.org/10.1002/sd.517.

Moulaert, Frank, Diane MacCallum, Abid Mehmood, and Abdelillah Hamdouch. 2013. "General Introduction: The Return of Social Innovation as a Scientific Concept and a Social Practice." In *The International Handbook on Social Innovation: Collective Action, Social Learning, and Transdisciplinary Research*, edited by Frank Moulaert, Diana MacCallum, Abid Mehmood, and Abdelillah Hamdouch, 1–6. Northampton, MA: Edward Elgar Publishing.

Mulgan, Geoff. 2007. *Social Innovation: What It Is, Why It Matters and How It Can Be Accelerated.* London: The Basingstoke Press.

Musitu-Ferrer, Daniel, Macarena Eseban-Ibanez, Celeste Leon-Moreno, and Oscar F. Garcia. 2019. "Is School Adjustment Related to Environmental Empathy and Connectedness to Nature?" *Psychosocial Intervention* 28 (2): 101–10.

Mutz, Michael, Johannes Muller, and Arne Goring. 2018. "Outdoor Adventures and Adolescents' Mental Health: Daily Screen Time as a Moderator of Changes." *Journal of Adventure Education and Outdoor Learning* 19 (1): 56–66.

Nash, Nick, Lorraine Whitmarsh, Stuart Capstick, Tom Hargreaves, Wouter Poortinga, Gregory Thomas, Elena Sautkina, and Dimitrios Xenias. 2017. "Climate-Behavioral Spillover and the Potential Contribution of Social Practice Theory." *Wires Climate Change* 8 (6). Accessed on July 2, 2022. 8:e481. doi: 10.1002/wcc.481.

National Climate Assessment. 2014. "Regions: Southeast and the Caribbean." *National Climate Assessment*. Washington, D.C. Accessed January 17, 2019. https://nca2014.globalchange.gov/report/regions/southeast.

National Weather Service: National Oceanic and Atmospheric Administration (NOAA). 2016. "Hurricane Mathew in the Carolinas, October 8, 2016." *National Weather Service.* October 8, 2016. Accessed January 2, 2019. https://www.weather.gov/ilm/Matthew.

Newton, Peter, and Denny Meyer. 2013. "Exploring the Attitudes-Action Gap in Household Resource Consumption: Does 'Environmental Lifestyle' Segmentation Align with Consumer Behavior?" *Sustainability* 5 (3): 1211–33. https://doi.org/10.3390/su5031211.

Nielsen, Thomas A. Sick. 2013. "Urban Planning Practices for Bikeable Cities—the Case of Copenhagen." *Urban Research & Practice* 6 (1): 110–15.

Nightingale, Andrea J. 2012. "The Embodiment of Nature: Fishing, Emotion, and the Politics of Environmental Values." In *Human-Environment Relations:*

Transformative Values in Theory and Practice, edited by Emily Brady and Pauline Phemister, 135–48. New York: Springer.

Nikitas, Alexandros. 2018. "Understanding Bike-Sharing Acceptability and Expected Usage Patterns in the Context of a Small City Novel to the Concept: A Story of 'Greek Drama.'" *Transportation Research Part F: Traffic Psychology and Behavior* 56: 306–21.

Nilsson, Andreas, Magnus Bergquist, and Wesley P. Schultz. 2017. "Spillover Effects in Environmental Behaviors, Across Time, and Context: A Review and Research Agenda." *Environmental Education Research* 23 (4): 573–89. https://doi.org/10.1080/13504622.2016.1250148.

Noblet, Caroline L., and John Thogersen. 2012. "Does Green Consumerism Increase the Acceptance of Wind Power?" *Energy Policy* 51: 854–62.

Nordback, Krista, Wesley E. Marshall, and Bruce N. Janson. 2014. "Bicyclist Safety Performance Functions for a U.S. City." *Accident Analysis & Prevention* 65: 114–22.

Nussbaum, Martha C. 2001. *Creating Capabilities: The Human Development Approach.* Cambridge: Harvard University Press.

O'Brien, Catherine. 2013. "Happiness and Sustainability Together at Last! Sustainable Happiness." *Canadian Journal of Education* 36 (4): 228–56.

Odell, Jenny. 2019. *How to Do Nothing: Resisting the Attention Economy.* Brooklyn: Melville House.

Olson, Erik L. 2012. "It's Not Easy Being Green: The Effects of Attribute Tradeoffs on Green Product Preference and Choice." *Journal of the Academy of Marketing Science* 41 (2): 171–84. https://doi.org/10.1007/s11747-012-0305-6.

Olsson, Per, Michele-Lee Moore, Frances R. Westley, and Daniel P. McCarthy. 2017. "The Concept of the Anthropocene as a Game-Changer: A New Context for Social Innovation and Transformations to Sustainability." *Ecology and Society* 22 (2): 31. https://doi.org/10.5751/ES-09310-220231.

O'Neill, Kirstie J., Adrian K. Clear, and Mike Hazas. 2019. "'Fractures' in Food Practices: Exploring Transitions Towards Sustainable Food." *Agriculture and Human Values* 36 (2): 225–39. https://doi.org/10.1007/s10460-019-09913-6.

Oriel, Sullivan, and Jonathan Gershuny. 2003. "Cross-national Changes in Time-use: Some Sociological (Hi)stories Re-examined." *The British Journal of Sociology* 52 (2): 331–47.

O'Shaughnessy, Sara, and Emily Huddart Kennedy. 2010. "Relational Activism: Reimagining Women's Environmental Work as Cultural Change." *Canadian Journal of Sociology* 35 (4): 551–72. https://www.jstor.org/stable/canajsocicahican.35.issue-4.

Overcash, Philip. 2009. Data from multiple email correspondences with Philip Overcash of the Department of Planning and Neighborhoods for the City of Charleston. October 21–December 11, 2009.

Pahl, Sabine, and Judith Bauer. 2013. "Overcoming the Distance: Perspective Taking with Future Humans Improves Environmental Engagement." *Environment and Behavior* 45 (2): 155–69. https://doi.org/10.1177/0013916511417618.

Palmetto Cycling Coalition and the South Carolina Livable Communities Alliance. 2018. "South Carolina and Pedestrian Crash Analysis 2009–2017." *Submitted by Equitable Cities, LLC.* Accessed on March 22, 2022. https://equitablecities.com/wp-content/uploads/2020/09/SC_Bicycle-and-Pedestrian-Data-Analysis_Report_Final_12.14.pdf.

Panter, Jenna R., Andrew P. Jones, and Esther M. van Sluijs. 2008. "Environmental Determinants of Active Travel in Youth: A Review and Framework for Future Research." *International Journal for Behavioral Nutrition and Physical Activity* 5 (34). https://doi.org/10.1186/1479-5868-5-34.

Parker, Julia, D., and Maureen H. McDonough. 1999. "Environmentalism of African Americans: An Analysis of the Subculture and Barriers Theories." *Environment and Behavior* 31 (2): 155–77.

Parker, Brenda, and Oona Morrow. 2017. "Urban Homesteading and Intensive Mothering: (Re)gendering Care and Environmental Responsibility in Boston and Chicago." *Gender, Place and Culture: A Journal of Feminist Geography* 24 (2): 247–59. https://doi.org/10.1080/0966369X.2016.1277186.

Patricio, Luis Claudio Brito, and Leandro Kruszielski. 2016. "Bike to Work Day: A Potential Tool for Planning and Fostering Sustainable Mobility." *Revista de Gestao Ambiental e Sustentabilidade* 5 (3): 135–45.

Payne, Phillip. 2006. "Embodiment and Environmental Education." *Environmental Education Research* 3 (2): 133–53.

Perkins, Tracy E. 2022. *Evolution of a Movement: Four Decades of California Environmental Justice Activism.* Oakland: University of California Press.

Pew Research Center. 2015. "Childlessness Falls, Family Size Grows Among Highly Educated Women." *Pew Research Center*. Accessed on March 25, 2022. https://www.pewresearch.org/social-trends/2015/05/07/family-size-among-mothers/

Pirchio, Sabine, Blanca Silvia Fraijo Sing, and Ylenia Passiatore, eds. 2020. "Special Issue: Where to Raise Happy and Skilled Children: How Environment Shapes Human Development and Education." *Frontiers in Psychology* 30.

Polanski, Nathan. 2019. "Reimagining the Right-of-Way: Designing Streets as Environmental and Economic Assets for Active Living and Healthy Environments." *Transportation Research Record: Journal of the Transportation Research Board* 2521: 139–48.

Powers, Hanna. 2021. "A Deadly Cycle: The Dangers of Biking in the Lowcountry." *News 2*. February 11, 2021. Accessed on March 22, 2022. https://www.counton2.com/news/local-news/charleston-county-news/a-deadly-cycle-the-dangers-of-biking-in-the-lowcountry/.

Prinzing, Michael. 2020. "Going Green Is Good for You: Why We Need to Change the Way We Think about Pro-Environmental Behavior." *Ethics, Policy & Environment* 2. Accessed on July 3, 2022. https://doi.org/10.1080/21550085.2020.1848192

Prothero, Andrea, Pierre McDonagh, and Susan Dobscha. 2010. "Is Green the New Black? Reflections on a Green Commodity Discourse." Journal of *MacroMarketing* 30 (2): 147–59. https://doi:10.1177/0276146710361922

Pucher, John, Ralph Buehler, and Mark Seinen. 2011. "Bicycling Renaissance in North America? An Update and Re-Appraisal of Cycling Trends and Policies."

Transportation Research Part A: Policy and Practice 45 (6): 451–75. https://doi.org/10.1016/j.tra.2011.03.001.

Racioppi, Francesca, Carlos Dora, and Harry Rutter. 2008. "Urban Settings and Opportunities for Healthy Lifestyles: Rediscovering Walking and Cycling and Understanding Their Health Benefit." *Built Environment* 31 (4): 302–14. https://doi.org/10.2148/benv.2005.31.4.302.

Reckwitz, Andreas. 2002. "Toward a Theory of Social Practices: A Development in Culturalist Theorizing." *European Journal of Social Theory* 5 (2): 243–63. https://doi.org/10.1177/13684310222225432.

Rector, Josiah. 2022. *Toxic Debt: An Environmental Justice History of Detroit (Justice, Power, and Politics).* Chapel Hill: The University of North Carolina Press.

Reid, Carlton. 2017. *Bike Boom: The Unexpected Resurgence of Cycling*. Washington, DC: Island Press.

Rerat, Patrick. 2021. "The Rise of the e-bike: Towards an Extension of the Practice of Cycling?" *Mobilities* 16 (3) June: 423–39. https://doi.org/10.1080/17450101.2021.1897236.

Rich, Stacey A., Sharon Hanna, and Bradley J. Wright. 2017. "Simply Satisfied: The Role of Psychological Need Satisfaction in the Life Satisfaction of Voluntary Simplifiers." *Journal of Happiness Studies* 18: 89–105.

Rindge, Brenda. 2016. "North Charleston to End Curbside Recycling in Dorchester Section of City." *Post and Courier*. July 24, 2016. Accessed June 24, 2022. https://www.postandcourier.com/archives/north-charleston-to-end-curbside-recycling-in-dorchester-section-of-city/article_c4ecb777-d69b-5cd7-8438-6fe66a58a598.html.

Ripple, William, J., Christopher Wolf, Thomas M. Newsome, Mauro Galetti, Mohammed Alamgir, Eileen Crist, Mahmoud I. Mahmoud, and William F. Laurance. 2017. "World Scientists' Warning to Humanity: A Second Notice." *BioScience* 67 (12): 1026–28. https://doi.org/10.1093/biosci/bix125.

Robertson, Margaret, and Ruth Lawrence, eds. 2015. *Experiencing the Outdoors: Enhancing Strategies for Wellbeing.* Boston: Brill.

Robinson, Jennifer Meta. 2016. "Making the Land Connection: Local Food Farms and Sustainability of Place." In *The Greening of Everyday Life: Challenging Practices, Imagining Possibilities*, edited by John M. Meyer and Jens M. Kerstin, 198–210. Oxford: Oxford University Press.

Rosa, Claudio D., Chistiana Cabicieri Profice, and Silvia Collado. 2018. "Nature Experiences and Adults' Self-Reported Pro-Environmental Behaviors: The Role of Connectedness to Nature and Childhood Nature Experiences." *Frontiers in Psychology* 9. Accessed on July 1, 2022. https://doi.org/10.3389/fpsyg.2018.01055

Rose, Eliot, Frank Gallivan, and Leslie Lara. 2016. "Survey-Based Evaluation of Bike-to-Work Day in the San Francisco Bay Area." *TRID: The TRIS and ITRD Database* No. 16–3852. Accessed on July 3. https://trid.trb.org/view/1393425

Roth, Wolff-Michael, and G. Michael Bowen. 2001. "Of Disciplined Minds and Disciplined Bodies: On Becoming an Ecologist." *Qualitative Sociology* 24 (4): 459–81. https://www.academia.edu/1377576/.

Rubens, Lolita, Patrick Gosling, Marino Bonaiuto, Xavier Brisbois, and Anne Moch. 2015. "Being a Hypocrite or Committed While I Am Shopping? A Comparison of the

Impact of Two Interventions on Environmentally Friendly Behavior." *Environment and Behavior* 47 (1): 3–16. https://doi.org/10.1177/0013916513482838.

Sahakian, Marlyne, and Harold Wilhite. 2014. "Making Practice Theory Practicable: Towards More Sustainable Forms of Consumption." *Journal of Consumer Culture* 14 (1): 25–44.

Scerri, Andy, and Liam Magee. 2012. "Green Householders, Stakeholder Citizenship and Sustainability." *Environmental Politics* 21 (3): 387–411. https://doi.org/10.1080/09644016.2012.671571.

Schlosberg, David, and Romand Coles. 2016. "The New Environmentalism of Everyday Life: Sustainability, Material Flows and Movements." In *The Greening of Everyday Life: Challenging Practices, Imagining Possibilities*, edited by John M. Meyer and Jens M. Kersten, 13–30. Oxford: Oxford University Press.

Schmidt, Luisa, Ana Horta, Augusta Correia, and Susana Fonseca. 2014. "Generational Gaps and Paradoxes Regarding Electricity Consumption and Saving." *Nature and Culture* 9 (2): 183–203. https://doi.org/10.3167/nc.2014.090205.

Schneider, Joachim, and Steffen Schaal. 2017. "Location-Based Smartphone Games in the Context of Environmental Education and Education for Sustainable Development: Fostering Connectedness to Nature with Geogames." *Environmental Education Research* 24 (11): 1597–1610.

Schor, Juliet B., and Craig J. Thompson, eds. 2014. *Sustainable Lifestyles and the Quest for Plenitude: Case Studies of the New Economy*. New Haven, CT: Yale University Press.

Schor, Juliet B., and Connor J. Fitzmaurice. 2015. Chapter 26, Collaborating and Connecting: the Emergence of the Sharing Economy." In *Handbook of Research on Sustainable Consumption*, edited by L. Reisch and J. Thogersen, 410–25. Northampton: Edward Elgar Publishing.

Schubert, Cornelius. 2014. *Social Innovations. Highly Reflexive and Multi-Referential Phenomena of Today's Innovation Society?* Berlin: Technical University, Technology Studies, Working Papers.

Scopelliti, Massimiliano, Daniela Barni, and Elena Rinallo. 2022. "My Parents Taught . . . Green Was My Growth! The Role of Intergenerational Transmission of Ecological Values in Young Adults' Pro-Environmental Behaviors and Their Psychosocial Mechanisms." *Environmental Research and Public Health* 19 (3): 1670. https://doi.org/10.3390/ijerph19031670.

Scott, James C. 1984. *Weapons of the Weak: Everyday Forms of Peasant Resistance*. New Haven, CT: Yale University Press.

Scurry, Ivy, and Vanessa Bowden. 2021. "'The Revolution's Never Done': The Role of 'Radical Imagery' within Anti-Capitalist Environmental Justice Activism." *Environmental Sociology* 7 (4): 316–26. https://doi.org/10.1080/23251042.2021.1916142.

Shea, Shannon Brescher. 2020. *Growing Sustainable Together: Practical Resources for Raising Kind, Engaged, Resilient Children*. Berkeley, CA: North Atlantic Books.

Shellenberger, Michael, and Ted Nordhaus. 2005. *The Death of Environmentalism: Global Warming Politics in a Post-Environmental World.* Accessed June 24, 2022. https://grist.org/article/doe-reprint/.

Shilling, Chris. 2003. *The Body and Social Theory*. London: Sage.

Shilling, Chris, and Philip A. Mellor. 2007. "Cultures of Embodied Experience: Technology, Religion and Body Pedagogics." *The Sociological Review* 55 (3): 531–49.

Shilling, Chris. 2004. "Physical Capital and Situated Action: A New Direction for Corporeal Sociology." *British Journal of Sociology of Education* 25 (4): 473–87.

Shipperijn, Jasper. 2017. "A Northern European Perspective on Creating More Activity Friendly Cities." *Preventive Medicine* 103: 53–54.

Shirani, Fiona, Catherine Butler, Karen Henwood, Karen Parkhill, and Nick Pidgeon. 2015. "'I'm not a Tree Hugger: I'm just like you': Changing Perceptions of Sustainable Lifestyles." *Environmental Politics* 24 (1): 57–74.

Shirani, Fiona, Catherine Anna Butler, Karen Henwood, Karen Parkhill, and Nicholas Frank Pidgeon. 2013. "Disconnected Futures: Exploring Notions of Ethical Responsibility in Energy Practices." *Local Environment* 18 (4): 455–68. https://doi.org/10.1080/13549839.2013.779236.

Shirani, Fiona, Christopher Groves, Catherine Butler, Karen Parkhill, Karen Henwood, and Nicholas Pidgeon. 2016. "Living the Future: Environmental Concerns, Parenting, and Low-Impact Lifestyles." In *Geographies of Global Issues: Change and Threat*, edited by N. Ansell, N. Klocker, and T. Skelton, 441–62. Singapore: Springer.

Shove, Elizabeth. 2010. "Beyond the ABC: Climate Change Policy and Theories of Social Change." *Environment and Planning A: Economy and Space* 42 (6): 1273–85. https://doi.org/10.1068/a42282.

Shove, Elizabeth, Mika Pantzar, and Matt Watson. 2012. *The Dynamics of Social Practice: Everyday Life and How It Changes*. London: Sage Publications.

Shyang-Chyuan, Fang. 2020/2021. "The Pro-Environmental Behavior Patterns of College Students Adapting to Climate Change." *Journal of Baltic Science Education* 5: 700–715.

Shwom, Rachael, and Janet A. Lorenzen. 2012. "Changing Household Consumption to Address Climate Change: Social Scientific Insights and Challenges." *WIREs Climate Change* 3 (5): 379–95. https://doi.org/10.1002/wcc.182.

Siegel, Lisa, Amy Cutter-Makenzie-Knowles, and Anne Bellert. 2018. "Still 'Minding the Gap' Sixteen Years Later: (Re)storying Pro-Environmental Behaviour." *Australian Journal of Environmental Education* 34 (2): 189–203. https://doi.org/10.1017/aee.2018.32.

Skinner, David, and Paul Rosen. 2007. "Hell Is Other Cyclists: Rethinking Transport and Identity." In *Cycling and Society*, edited by Dave Horton, Paul Rosen, and Peter Cox, 83–96. New York: Routledge.

Slade, David. 2018. "105,000 Homes Await Construction in the Charleston Metro Area." *Post and Courier*. July 18, 2018. Accessed January 2, 2019. https://www.postandcourier.com/news/105-000-homes-await-construction-in-the-charleston-metro-area/article_e44b5830-86ef-11e8-9f62-6be4909033c5.html.

Soper, Kate. 2007. "Re-thinking the 'Good Life': The Citizenship Dimension of Consumer Disaffection with Consumerism." *Journal of Consumer Culture* 7 (2): 205–29.

Soper, Kate, Martin H. Ryle, and Thomas Lyn. 2009. *Politics and Pleasures of Consuming Differently*. London: Palgrave Macmillan.

Soper, Kate. 2011. "Alternative Hedonism, Cultural Theory and the Role of Aesthetic Revisioning." In *Cultural Studies and Anti-Consumerism: A Critical Encounter*, edited by S. Binkley and J. Littler, 49–69. New York: Routledge.

Southern Cross University. 2019. Going Green (blog). July 7, 2019. Accessed February 26, 2022. https://online.scu.edu.au/blog/going-green/.

Speck, Melanie, and Marco Hasselkuss. 2015. "Sufficiency in Social Practice: Searching Potentials for Sufficient Behavior in a Consumerist Culture." *Sustainability: Science Practice and Policy* 11 (2): 1–19. https://doi.org/10.1080/15487733.2015.11908143.

Spijker, Stephanie Nuria, Erik Mathijs, and Constanza Parra. 2020. "Grasping Practices of Self-Reliance within Alternative Foodscapes in Flanders." *Agriculture and Human Values* 37 (3): 819–32. https://doi.org/10.1007/s10460-020-10021-z.

Spotswood, Fiona, Tim Chatterton, Alan Tapp, and David Williams. 2015. "Analysing Cycling as a Social Practice: An Empirical Grounding for Behaviour Change." *Transportation Research Part F: Traffic Psychology and Behaviour* 29: 22–33. https://doi.org/10.1016/j.trf.2014.12.001.

Sterling, Stephen. 2011. "Transformative Learning and Sustainability: Sketching the Conceptual Ground." *Learning and Teaching in Higher Education* 5: 17–33.

Stern, Paul C. 2000. "New Environmental Theories: Toward a Coherent Theory of Environmentally Significant Behavior." *Journal of Social Issues* 56 (3): 407–24. https://doi.org/10.1111/0022-4537.00175.

Stevenson, Kathryn, T. 2016. "The Influence of Personal Beliefs, Friends, and Family in Building Climate Change Concern Among Adolescents." *Environmental Education Research* 25 (6): 832–45.

Stock, Paul V., and Jeremie Forney. 2014. "Farmer Autonomy and the Farming Self." *Journal of Rural Studies* 36: 160–71.

Stolle, Dietlind, and Michelle Micheletti. 2013. *Political Consumerism: Global Responsibility in Action.* Cambridge: Cambridge University Press.

Stoner, Krystina R., Michael A. Tarrant, Lane Perry, Lee Stoner, Stephen Wearing, and Kevin Lyons. 2014. "Global Citizenship as a Learning Outcome of Educational Travel." *Journal of Teaching in Travel & Tourism* 14 (2): 149–63. https://doi.org/10.1080/15313220.2014.907956.

Strengers, Yolande, and Cecily Maller, eds. 2015. *Social Practices, Intervention, and Sustainability: Beyond Behaviour Change*. New York: Routledge.

Stuart, Diana, Ryan Gunderson, and Brian Petersen. 2020. "Overconsumption as Ideology: Implications for Addressing Global Climate Change." *Nature and Culture* 15 (2): 199–223.

Sussman, Reuven, and Robert Gifford. 2013. "Be the Change You Want to See." *Environment and Behavior* 45 (3): 323–43.

Sutton, Anna. 2020. "Living the Good Life: A Meta-Analyis of Authenticity, Well-Being, and Engagement." *Personality and Individual Differences* 153. Accessed on June 14, 2022. https://doi.org/10.1016/j.paid.2019.109645

Szasz, Andrew. 2007. *Shopping Our Way to Safety: How We Changed from Protecting the Environment to Protecting Ourselves.* Minneapolis: University of Minnesota Press.

Tarasoff, Lesley Ann. 2018. "A Qualitative Study of Embodiment among Women with Physical Disabilities during the Perinatal and Early Motherhood." PhD diss. University of Toronto. Accessed on June 28, 2022. https://tspace.library.utoronto.ca/bitstream/1807/92020/1/Tarasoff_Lesley_A_201811_PhD_thesis.pdf

Tarrant, Michael A., Lee Stoner, William T. Borrie, Gerald Kyle, Roger L. Moore, and Annette Moore. 2017. "Educational Travel and Global Citizenship." *Journal of Leisure Research* 43 (3): 403–26. https://doi.org/10.1080/00222216.2011.11950243.

Taylor, Dorceta, E. 2014. *Toxic Communities: Environmental Racism, Industrial Pollution, and Residential Mobility.* New York: New York University Press.

Teixido-Figueras, Jordi, Julia K. Steinberger, Fridolin Krausmann, Helmut Haberl, Thomas Wiedmann, Glen P. Peters, Juan A. Duro, and Thomas Kastner. 2016. "International Inequality of Environmental Pressures: Decomposition and Comparative Analysis." *Ecological Indicators* 62: 163–73. https://doi.org/10.1016/j.ecolind.2015.11.041.

Thogersen, John, and Folke Olander. 2003. "Spillover of Environment-Friendly Consumer Behavior." *Journal of Environmental Psychology* 23 (3): 225–36. https://doi.org/10.1016/S0272-4944(03)00018-5.

Thomas, Gregory O., Rose Fisher, Lorraine Whitmarsh, Taciano L. Milfont, and Wouter Poortinga. 2018. "The Impact of Parenthood on Environmental Attitudes and Behaviour: A Longitudinal Investigation of the Legacy Hypothesis." *Population and Environment* 39 (3): 261–76. https://doi.org/10.1007/s11111-017-0291-1.

Tiefenbeck, Verena, Thorsten Staake, Kurt Roth, and Olga Sachs. 2013. "For Better or for Worse? Empirical Evidence of Moral Licensing in a Behavioral Energy Conservation Campaign." *Energy Policy* 57 (C): 160–71. https://doi.org/10.1016/j.enpol.2013.01.021.

Tillmann, Suzanne, Danielle Tobin, William Avison, and Jason Gilliland. 2018. "Mental Health Benefits of Interactions with Nature in Children and Teenagers: A Systematic Review." *Epidemiology & Community Health* 72 (10): 958–66. Accessed on July 2, 2022. http://dx.doi.org/10.1136/jech-2018-210436

Tsai, Chin-Chang, Li (Dale) Xiangyu, and Wu Wei-Ning. 2021. "Explaining Citizens' Pro-Environmental Behaviors in Public and Private Spheres: The Mediating Role of Willingness to Sacrifice for the Environment." *Australian Journal of Public Administration* 80 (3): 510–39.

Twine, Richard, T. 2001. "Ma(r)king Essence-Ecofeminism and Embodiment." *Ethics and the Environment* 6 (2): 31–58.

Tyers, Roger, Tristan Berchoux, and Kun Xiang. 2018. "China-to-UK Student Migration and Pro-Environmental Behaviour Change: A Social Practice

Perspective." *Sociological Research Online* 24 (4): 575–97. https://doi.org/10.1177%2F1360780418794194

UCS (Union of Concerned Scientists). 2017. "Historic Communities Face New Challenges as Sea Levels Rise along the South Carolina Coast." *Union of Concerned Scientists.* June 26, 2017. Accessed June 5, 2019. https://www.ucsusa.org/resources/historic-communities-face-new-challenges-sea-levels-rise-along-south-carolina-coast

Ulug, Ciska, Elen-Maarja Trell, and Lummina Horlings. 2021. "Ecovillage Foodscapes: Zooming In and Out of Sustainable Food Practices." *Agriculture and Human Values* 38 (4): 1041–59. https://doi.org/10.1007/s10460-021-10213-1.

UNEP (United Nations Environment Programme). 2010. *Task Force on Sustainable Lifestyles*. Accessed May 15, 2019. https://www.strategicdesignscenarios.net/wp-content/uploads/2012/06/UNEP_SLT-Report.pdf

UNEP (United Nations Environment Programme). 2021a. *Annual Report: Planetary Action. United Nations*. Accessed on March 20, 2022. https://www.unep.org/annualreport/2021/index.php.

UNEP (United Nations Environment Programme). 2021b. *Greening the Blue Report 2021: The UN System's Environmental Footprint and Efforts to Reduce It.* Accessed on June 29, 2022. https://wedocs.unep.org/handle/20.500.11822/37330

Urban, Mark C. 2015. "Accelerating Extinction Risk from Climate Change." *Climate Change* 348 (6234): 571–73. https://doi.org/10.1126/science.aaa4984.

U.S. Census Bureau. 2022a. "QuickFacts: Charleston County, South Carolina." *Census.gov Quick Facts*. Accessed March 22, 2022. https://www.census.gov/quickfacts/charlestoncountysouthcarolina.

U.S. Census Bureau. 2022b. "QuickFacts: United States. Accessed March 22, 2022. https://www.census.gov/quickfacts/fact/table/US/PST045221.

U.S. Census Bureau. 2016. "A Look at the Nearly 1 Million Who Ride Their Bikes to Work in the U.S." *Census.gov Newsroom*. Accessed April 15, 2022. https://www.census.gov/newsroom/blogs/random-samplings/2016/05/a-look-at-the-nearly-1-million-who-ride-their-bikes-to-work-in-the-u-s.html#:~:text=The%20number%20of%20men%20who,with%2072%20percent%20in%202014.

U.S. Department of Commerce. 2016. "Service Assessment: The Historic South Carolina Floods of October 1–5, 2015." *National Oceanic Atmospheric Administration, National Weather Service,* Silver Spring, MD. Accessed December 23, 2018. https://www.weather.gov/media/publications/assessments/SCFlooding_072216_Signed_Final.pdf.

U.S. DOT (US Department of Transportation). 2015. "Commute Mode Share." Accessed on April 15, 2022. https://www.transportation.gov/mission/health/commute-mode-share.

US News and World Report. 2022. "Fastest-Growing Places in the U.S. in 2022–2023." Accessed March 22, 2022. https://realestate.usnews.com/places/rankings/fastest-growing-places?src=usn_mayoroffice.

Uyeki, Eugene S., and Lani J. Holland. 2000. "Diffusion of Pro-Environmental Attitudes?" *American Behavioral Scientist* 43: 646–62.

Van Dam, Ynte K., and Arnout R. H. Fischer. 2015. "Buying Green Without Being Seen." *Environment and Behavior* 47 (3): 328–56. https://doi.org/10.1177/0013916513509481.

Vandenbergh, Michael P., Paul C. Stern, Gerald T. Gardner, Thomas Dietz, and Jonathan M. Gilligan. 2011. "Implementing the Behavioral Wedge: Designing and Adopting Effective Carbon Emissions Reduction Programs." *Environmental Forum* 28 (4): 54–63.

Verplanken, Bas, Ian Walker, Adrian Davis and Michaela Jurasek. 2008. "Context, Change, and Travel Mode Choice: Combining the Habit Discontinuity and Self-Activation Hypotheses." *Journal of Environmental Psychology* 28 (2): 121–27.

Verplanken, Darnton A., Piron White, and Louise Whitmarsh. 2011. *Habits, Routines, and Sustainable Lifestyles: A Summary Report to the Department for the Environment, Food, and Rural Affairs.* London: AD Research and Analysis for Defra.

Wacquant, Loic. 2015. "For a Sociology of Flesh and Blood." *Qualitative Sociology* 38: 1–11.

Wacquant, Loic. 2006. *Body and Soul: Notebooks of an Apprentice Boxer*. Oxford: Oxford University Press.

Wacquant, Loic. 2005. "Carnal Connections: On Embodiment, Apprenticeship, and Membership." *Qualitative Sociology* 28: 445–74.

Wajcman, Judy, and Nigel Dodd. 2017. *The Sociology of Speed: Digital, Organizational, and Social Temporalities*. Oxford: Oxford University Press.

Walther, Carol S., Jennifer A. Sandlin, and Kristi Wuensch. 2016. "Voluntary Simplifiers, Spirituality, and Happiness." *Humanity & Society* 40 (1): 22–42.

Washburn, Rachel. 2014. "Measuring Personal Chemical Exposures through Biomonitoring: The Experiences of Research Participants." *Qualitative Health Research* 24 (3): 329–44. https://doi.org/10.1177/1049732314521899.

Webb, Janette, David Hawkey, David McCrone, and Margaret Tingey. 2016. "House, Income, and Transforming Energy in a Cold Climate." *Climate Change, Sustainability, and Biodiversity* 5 (3): 411–29.

Weidmann, Thomas, Manfred Lenzen, Lorenz T. KeyBer, and Julia K. Steingraber. 2020. "Scientists' Warning on Affluence." *Nature Communications* 11: 1–10. Accessed March 19, 2022. https://www.nature.com/articles/s41467-020-16941-y.pdf.

Weiss, Gail, and Honi Fern Haber, eds. 1999. *Perspectives on Embodiment: The Intersections of Nature and Culture.* London: Routledge.

Weiss, Robert S. 1994. *Learning From Strangers: The Art and Method of Qualitative Interview Studies*. New York: Free Press.

Wells, Nancy M., and Kristi S. Lekies. 2006. "Nature and the Life Course: Pathways from Childhood Nature Experiences to Adult Environmentalism." *Children, Youth and Environments* 16 (1): 1–24.

Wells, Nancy M., and Kristi S. Lekies. 2012. "Chapter 14, Children and Nature: Following the Trail to Environmental Attitudes and Behavior." In *Citizen Science:*

Public Participation in Environmental Research, edited by J. Dickinson and R. Bonney, 201–13. Ithaca: Cornell University Press.

Werfel, Seth H. 2017. "Household Behavior Crowds Out Support for Climate Change Policy When Sufficient Progress Is Perceived." *Nature Climate Change* 7 (7): 512–15.

White, Monica. 2011. "Sisters of the Soil: Urban Gardening as Resistance in Detroit." *Race/Ethnicity: Multidisciplinary Global Contexts. Food Justice* 5 (1): 13–28.

Whitman, Gail, and William H. Cooper. 2000. "Ecological Embeddedness." *The Academy of Management Journal* 43 (6): 1265–82.

Whitman, Gail, and William H. Cooper. 2011. "Ecological Sensemaking." *The Academy of Management Journal* 54 (5): 889–911.

Whitmarsh, Lorraine. 2022. "Leveraging Moments of Change for Pro-Environmental Behavioral Transformation." Podcast from the London School of Economics and Political Science. February 11, 2022. Accessed March 3, 2022. https://www.lse.ac.uk/lse-player?id=c759f46f-64a9-4511-b228-2c05fe1f7c2e.

Wilhite, Hal. 2015. "The Problem of Habits for a Sustainable Transformation." In *Sustainable Consumption and the Good Life: Interdisciplinary Perspectives,* edited by Karen L. Syse and Martin Lee Mueller, 100–110. New York: Routledge.

Wilhoit, Elizabeth D., and Lorraine G. Kisselburgh. 2017. "The Relational Ontology of Resistance: Hybridity, Ventriloquism, and Materiality in the Production of Bike Commuting as Resistance." *Interdisciplinary Journal of Organization, Theory and Society* 26 (6): 873–93. https://doi.org/10.1177/1350508417723719.

Williams, Corrie Colvin, and Louise Chawla. 2016. "Environmental Identity Formation in Nonformal Environmental Education Programs." *Environmental Education Research* 22 (7): 978–1001.

Williamson, Katie, Aven Satre-Meloy, Katie Velasco, and Kevin Green. 2018. "Climate Change Needs Behavior Change: Making the Case for Behavioral Solutions to Reduce Global Warming." *Center for Behavior & the Environment*. Arlington, VA: Rare. https://docplayer.net/135908972-Climate-change-needs-behavior-change-making-the-case-for-behavioral-solutions-to-reduce-global-warming.html.

Willis, Margaret M., and Juliet B. Schor. 2012. "Does Changing a Light Bulb Lead to Changing the World? Political Action and the Conscious Consumer." *ANNALS of the American Academy of Political and Social Sciences* 644 (1): 160–90. https://doi.org/10.1177/0002716212454831.

Windhorst, Eric, and Allison Williams. 2015. "Growing Up, Naturally: The Mental Health Legacy of Early Nature Affiliation." *Ecopscychology* 7 (3): 115–25.

Wood, Carly J., and Nina S. Smyth. 2020. "The Health Impact of Nature Exposure and Green Exercise across the Life Course: A Pilot Study." *International Journal of Environmental Health Research* 30 (2): 226–35.

Woods, Lisa, Billie Giles-Corti, Stephen R. Zubrick, and Max K. Bulsara. 2013. "Through the Kids . . . We Connected With Our Community." *Environment and Behavior* 45 (3): 344–68. https://doi.org/10.1177/0013916511429329.

World Economic Forum. 2019. "The Global Risks Report: 2019." January 15, 2019. Accessed June 5, 2019. https://www.weforum.org/reports/the-global-risks-report-2019.

World Population Review. 2022. "Charleston, South Carolina Population 2022." Accessed March 22, 2022. https://worldpopulationreview.com/us-cities/charleston-sc-population

Wu, Tim. 2018. *Stand Out of Our Light: Freedom and Resistance in the Attention Economy*. Cambridge: Cambridge University Press.

Wu, Tim. 2016. *The Attention Merchants: The Epic Scramble to Get Inside Our Heads*. 2016. New York: Vintage Books.

Zelenski, John M., and Elizabeth K. Nisbet. 2014. "Happiness and Feeling Connected: The Distinct Role of Nature Relatedness." *Environment and Behavior* 46 (1): 3–23. https://doi.org/10.1177/0013916512451901.

Zhao, Pengjun. 2014. "The Impact of the Built Environment on Bicycle Commuting: Evidence from Beijing." *Urban Studies* 51 (5): 1019–37. https://doi.org/10.1177/0042098013494423.

Zheng, Qiujin, Yunjian Zheng, Qiuqin Zheng, and Xiaofeng Su. 2020. "Effects of Environmental Education and Environmental Facilities on Visitors' Environmental Literacy: A Case of Rural Tourism." *Revista de Cercetare Si Interventie Sociala* 69: 313–23. http://dx.doi.org/10.33788/rcis.69.20.

Zylstra, Matthew J., Andrew T. Knight, Karen J. Esler, and Lesley L. L. Le Grange. 2014: *Springer Science Reviews* 2: 119–43.

Appendix A

SEMI-STRUCTURED INTERVIEW GUIDE FOR GREEN PARENTING STUDY

1. Green Parenting and Green Living History
2. Daily Green Habits
3. Challenges and Opportunities
4. Sustainable Living and Choice of Career or Work
5. How to Help Others Live More Sustainably
6. Narrative (What It Means to Be a "Green" Parent)
7. Additional Comments
8. Demographics

Appendix B

SEMI-STRUCTURED INTERVIEW GUIDE FOR CHOICE COMMUTER STUDY

1. Bike Commuting History and Likes/Dislikes
2. Connections Between Bike Commuting and Ideological Beliefs
3. Driving Likes/Dislikes
4. Daily Bike Commuting Experiences
5. Facilitators/Barriers to Bike Commuting
6. Bike Commuting and Consumption Habits
7. Connections to Social Networks
8. Involvement in Advocacy
9. A Story Telling about Commuting
10. Additional Comments
11. Demographics

Appendix C

SEMI-STRUCTURED INTERVIEW GUIDE FOR NECESSITY COMMUTER STUDY

1. Bike Commuting History and Likes/Dislikes
2. Driving Likes/Dislikes
3. Daily Bike Commuting Experiences
4. Facilitators/Barriers to Bike Commuting
5. A Story Telling about Commuting
6. Additional Comments
7. Demographics

Index

About the Author

Deborah McCarthy Auriffeille, PhD, is associate professor of sociology and associate chair of sociology and anthropology at the College of Charleston. Deb Auriffeille currently studies environmental lifestyles with a focus on green parenting and bike commuting. She is also coeditor, with Leslie King, of the four editions of *Environmental Sociology: From Analysis to Action.* Most recently, Deb coauthored a magazine-style article on green parenting for the peer-reviewed journal *Contexts*, with Cat Fleming, a College of Charleston student researcher and sociology major.

www.ingramcontent.com/pod-product-compliance
Lightning Source LLC
LaVergne TN
LVHW091531200926
848483LV00002B/368